Technology and Society
in Twentieth Century America
An Anthology

Technology and Society in Twentieth Century America

An Anthology

Randall E. Stross
San Jose State University

Wadsworth Publishing Company
Belmont, California
A Division of Wadsworth, Inc.

Sponsoring editor: Casimir Psujek
Project editor: Mary Lou Murphy
Production manager: Ann Cassady
Compositor: Graphic World Inc.
Typeface: 10/12 Palatino
Printer: Malloy Lithographing, Inc.

Library of Congress Cataloging-in-Publication Data

Technology and society in twentieth century America.

    1.   Technology—Social aspects—United States.
I. Stross, Randall E.
T21.T429 1989   303.4'83   88–7162
ISBN 0-534-10927-6 (previously ISBN 0-256-06216-1)

Printed in the United States of America

    2 3 4 5 6 7 8 9 0 ML 5 4 3 2 1 0

Contents

INTRODUCTION

We live in an era in which change seems to be the only constant. Technological change, most visible of all, takes place at a breathtaking pace. Our world is remade again and again as our technical power to create, and to destroy, advances at a speed that accelerates with each passing decade. To make sense of the dizzying changes that have recently taken place, we can take stock of some of the most important technological developments of this century and the impact of these changes on our lives—an impact that is often overlooked in the course of going about our daily routines. Here, in the closing years of the twentieth century, we can look back at the phenomenon of rapid technological change and give thought to how technology has influenced the shape of American life.

Historians have no convenient label that neatly sums up the current age and a single dominant technology. "The Machine Age" has already been used to refer to the arrival of machines in large-scale factories in the post–Civil War decades. In the early twentieth century, American social life seemed to be dominated for brief periods by single inventions, giving rise then and later to catchphrases such as the "Age of Radio" or the "Age of the Automobile." More recently, we have often heard mention of possible successors: "the Atomic Age" or the "post–industrial Information Age." Today, however, it is difficult to claim that any one technology deserves to be called the defining technology of the times. We live today in a society that is permeated with the culminating influence of the many important technologies that have appeared over time, and in the cacophony, no single voice stands apart from the rest.

We are also more ambivalent toward technology than ever before; the earlier, almost total enthusiasm for technology is now co-mixed with some disillusionment. As recently as 1969, when the country celebrated the Apollo mission's landing on the moon, the achievement was seen in two lights; both as a fulfillment of the country's self-defined historical destiny to serve as a technological leader, and also as a symbol of our faith that there was literally no limit to what humans could do. Since that time, however, Americans have revised their view of

1

America as unquestioned leader, and of technology as a panacea for all problems. Perhaps more than anything else, increased awareness of technology's destructive impact on the natural environment has curbed our earlier, naive assumption that all technological advancement could be equated with progress and improvement.

Other events have reshaped American attitudes toward technology. Since the moon landings, the nation's space program has been beset with problems, and, as in the past, has been used to symbolize a larger American character. Now, instead of standing for the country's ability to deliver technological achievements, the space program and its woes stand for America's faltering ability to produce leading technology. Adding to the self-doubt has been the deterioration of the country's trade competitiveness and the concern that our trade problems stem, at least partially, from lagging technological innovation. Increased attention has been focused on our schools. Policymakers and the larger public wonder aloud how the educational system should be overhauled to overcome "technological illiteracy" and other problems contributing to the eroding competitiveness of American products in world markets.

In some of the selections in this book, you will hear a tone of self-doubt or ambivalence when the author discusses a particular aspect of technology's impact on society. In some selections, you will also find that the very definition of technology is broad. The most common view of technology is one that equates technology with inanimate objects— tools or machines, or, as we now are accustomed to saying, hardware. "Technology" refers to more than tools or machines. It also includes other things that are often overlooked, such as the human knowledge that permits us to accomplish particular tasks. The ability to design a computer, or, let us say, a microwave oven, is critically important to the production of these electronic boxes that adorn our offices and homes today. Technical knowledge, or "technique," is not visible—it is a resource that resides in humans and in books—yet it is just as properly regarded as a form of technology as are tools.

We should also realize that technology embraces the ways in which we organize ourselves and our knowledge to accomplish particular ends. For example, a number of different ways of transmitting knowledge to undergraduates are used. Classes range from small to large; prerequisites vary; so, too, do reading loads and dozens of other organizational details related to the delivery of information. Any of these various forms of teaching should be seen as different "technologies." Hardware does not have to be present—the lecture method of teaching is just as properly regarded as a kind of "technology" as is the use of microcomputers or overhead projectors. When we think about the importance of social organization in our accomplishment of particular ends, we can appreciate that animate beings—humans—are as bound to the very definition of technology as are inanimate tools. In fact, some scholars regard the organizational system that the ancient Egyptians used to build the pyramids to be the first "machine."

If we think of technology as both hardware and software, as both machines and the techniques humans use to direct themselves and their tools, then we can see a wide variety of possible topics related to technology that can be explored in twentieth-century America. For the most part, the historians and other writers who have written about technology in American history can be sorted into one of two groups. One group writes about technology's internal development—about who invented what and when. The other group writes about the external aspects of technology—of what have been the effects, intended and unintended, of new technology upon the wider society. This collection of readings on technology and society, by necessity, is selective, and represents the work of the latter group. It focuses on how technological developments, such as the automobile, the television, and the computer, have affected society, for better or for worse.

In reality, the relationship between technology and society is not so simple. Society, of course, affects the course of technological development and the choices that are made between alternative technical projects. But the history of technological invention has, for the most part, been excluded from this book in order to provide room for readings on technology's social impact, the aspect that has received comparatively less attention to date.

One other aim of this book should be mentioned, which is to try to remove the mysteriousness that has come to surround modern technology. We hear older Americans lament the passing of the time in which an ordinary household appliance, like a toaster, could be fixed by anyone with an interest in tinkering with things. Today, an appliance that breaks down may require replacement of semiconductor chips that, when inspected by an intrepid amateur, yield no visible clues concerning their operating condition. The chips are encased in black plastic that cannot be opened—an enigma that dampens natural curiosity. Although the sophistication of today's technology places much of it beyond the technical understanding of the general public, we can still try to dispel the air of mystery by paying more attention to technology's influence on our lives.

The selections that are presented here draw on recently published books. Besides historians, we hear from sociologists, journalists, and intellectual gadflies who defy simple categorization. Some of the authors argue that a given technology has had more of an impact on twentieth-century American life than any other; others go so far as to imply that technology is *the* most influential force in our history, a viewpoint that would be called "technological determinism." Technology's impact is celebrated in some cases, deplored in others. All of the views that are presented in this book, however, should be seen as the work of very human authors whose individual views may or may not seem persuasive to you, the reader. Their interpretations of how technology has influenced modern life are just that—interpretations. All of these selections invite discussion, challenge, and debate.

CHAPTER 1

The Americans: The Democratic Experience

Daniel J. Boorstin

If anyone would dare tackle the task of describing in one book the entire history of technology's impact on modern American society, it would be Daniel J. Boorstin, Librarian of Congress Emeritus. An omnivorous reader with the broadest of interests, Boorstin has always tackled large projects. This book is the third in a trilogy covering the sweep of American history from colonial times to the present.

The Democratic Experience concerns the years since the Civil War, and the selected portions presented here focus mainly on events that occurred in the late nineteenth or the early years of the twentieth century. In setting the stage for the readings that follow, the Boorstin selections show us how technological innovations transformed life in the city in many different and often overlooked ways.

In the first selection, "Consumers' Palaces," we look at the rise of department stores, which Boorstin reminds us rested on important technological developments. Cast iron, elevators, and plate glass made it possible for builders to construct high-ceilinged, multistoried, and bright store "palaces." Electric trolleys and cheap public transportation allowed people living in outlying areas to travel conveniently to the central urban districts to shop, and newspaper advertising helped lure them there.

Boorstin emphasizes an important theme here and throughout his work: Technology in America has been just as important in making our nation democratic as have our political institutions because technology has served to equalize opportunities and comforts. The new department stores, for example, were too large for the owners to entrust the many store employees with bargaining with customers over prices, so they adopted the fixed-price policy, making it possible for all customers to purchase an item for the same price, regardless of their varying skills in the fine art of haggling.

When one company opened many different stores around the country, these "chain stores" were able to achieve unprecedented economies of scale and offer their customers low prices. They also democratized experience, in the sense that customers in widely scattered locations purchased identical goods on a mass scale for the first time. Chain stores were bitterly opposed in many communities, but their expansion was unstoppable.

In the selection "Nationwide Customers," Boorstin refers in passing to an "anti–RFD movement" at the same time chain stores were expanding, and this deserves explanation. "RFD" refers to Rural Free Delivery. As late as 1900, only urban residents enjoyed free home mail delivery; farmers had to go to the nearest town to pick up their mail. In the 1890s, when farmer groups demanded RFD, they were opposed by those who regarded a rural mail delivery system as unconscionably expensive, but in the end, in 1906, the farmers finally got RFD.

The final selection, "Homogenizing Space," presents what we might call a "social history of the bathroom." The author shows us the slow pace at which cities established centralized water supply systems, and purification of tap water came even more slowly. Once residents were provided with convenient delivery of water, however, their consumption of the resource expanded much more quickly than anyone had predicted. (Later, our consumption of electricity would follow a similar pattern). Running water, and certain innovations in the technology of toilet fixtures, led to the establishment of separate rooms—"water closets," or bathrooms. These, Boorstin points out, accommodated only one person at a time and brought to an end a tradition of conversing with others while taking care of bodily functions.

Central heating, then air conditioning, led to "homogenization of space," or the ability to keep indoor temperature and humidity conditions constant throughout the year. As giant shopping malls and sports complexes have expanded the cubic area of what is technically "indoor" space, Boorstin calls our attention to a new phenomenon of "indoor-outdoor confusion," and introduces the Houston Astrodome's interesting history.

Boorstin describes the Astrodome's gift to modern civilization—the artificial grass called Astroturf—as being, for sports purposes, "superior to the real thing." Longer experience with synthetic turf has led many to question its alleged superiority, and some sports arena owners have gone to the trouble and expense of pulling up the artificial grass and sodding and seeding their fields. But there is a larger question that could be raised about artificial turf and about the other technological developments described in this reading: What *aesthetic* cost is paid for Astroturf? Or for air conditioning and our indoor-outdoor confusion?

Or, we might ask, what aesthetic cost has been paid for the "malling of America," as we often hear lamented today. The phenomenal proliferation of vast franchise empires and seemingly identical shopping malls we see today mirrors the onslaught of the chain stores of a hundred years earlier. Identical fast-food outlets provide a democratic experience accessible to residents virtually anywhere in the country, but the outlets also have "homogenized experience," making a visit to a restaurant the same regardless of region or local culinary traditions. Boorstin shows us that these issues are not new.

CONSUMERS' PALACES

Between the Civil War and the beginning of the new century there appeared grand and impressive edifices—Palaces of Consumption—in the principal cities of the nation and in the upstart cities that hoped to become great metropolises. A. T. Stewart's, Lord & Taylor, Arnold Constable, R. H. Macy's in New York City; John Wanamaker in Philadelphia; Jordan Marsh in Boston; Field, Leiter & Co. (later Marshall Field & Co.) and the Fair in Chicago. And even smaller cities had their impressive consumers' palaces—Lazarus in Columbus, Ohio, and Hudson's in Detroit, among others.

The distinctive institution which came to be called the department store was a large retail shop, centrally located in a city, doing a big volume of business, and offering a wide range of merchandise, including clothing for women and children, small household wares, and usually dry goods and home furnishings. While the stock was departmentalized, many of the operations and the general management were centralized. If the department store was not an American invention, it flourished here as nowhere else. "Department store" was an Americanism in general use before the opening of the twentieth century.

The grand new consumers' palaces were to the old small and intimate shops what the grand new American hotels were to the Old World inns. Like the hotels, the department stores were symbols of faith in the future of growing communities. For citizens of the sprouting towns the new department-store grandeur gave dignity, importance, and publicity to the acts of shopping and buying—new communal acts in a new America.

Alexander Turney Stewart, at the age of seventeen, came to New York City from Northern Ireland and began his business with a stock of Irish laces. Only fifteen years later, in 1846, he built an impressive struc-

SOURCE: From *The Americans: The Democratic Experience* by Daniel J. Boorstin. Copyright © 1973 by Daniel J. Boorstin. Reprinted by permission of Random House, Inc.

ture at Broadway and Chambers streets, known as the Marble Dry-Goods Palace. Like many another earlier palace, it expanded with addition after addition until it extended along a two-hundred-foot frontage on City Hall Park and covered the whole block on Broadway. In 1862, when Stewart's outgrew these premises, it moved into another palace—this time eight stories high and no longer of marble. This building, which became famous as Stewart's Cast Iron Palace, was reputed to be the largest retail store in the world.

The new department stores, unlike the elegant exclusive shops of Old World capitals, were palatial, public, and inviting. Cast iron made it easier than ever to make buildings impressive on the outside, and on the inside to offer high ceilings, and wide, unbroken expanses for appealing display. In the five-story E. V. Haughwout Department Store, built in 1857 at Broadway and Broome Street in New York City, Daniel Badger, pioneer in manufacturing iron for buildings, offered his most impressive work. The intricate façades of the Venetian *palazzos* were easily reproduced in cast iron. Their elegant patterns of columns, spandrels, and windows could be endlessly extended around a building, and the architectural orders could be piled one above another indefinitely.

When James Bogardus (the prolific inventor whose works included a metal-cased pencil with a lead "forever pointed," improvements in the striking parts of clocks, a new machine for making postage stamps, and an improved mill for making lead paint) turned his genius to finding new uses for cast iron, the needs of the department store excited his imagination. These new iron structures, he exulted, could be raised to a height of ten miles. Bogardus would exploit qualities in the cast-iron frame—lightness, openness, adaptability, and speed of construction—similar to those which three decades before had given the balloon frame its special American appeal.

The climax of this new Iron Age was the Cast Iron Palace which Bogardus built on Broadway between Ninth and Tenth streets for A. T. Stewart. It was the largest iron building of its day, one of the largest of any kind. On the exterior, the molded iron panels were painted to resemble stone; the repeating column-and-beam design added dignity and expansiveness. Each floor took the weight of its own outer walls, in the structural scheme which would make possible the skyscraper. The thin walls at the ground floor produced a spacious, open lobby, and the slender iron columns kept vistas open on every floor, vistas of appealing merchandise of all shapes, color, and description, objects one had never thought of seeing, much less of buying. And one could see out there among the merchandise the enticing crowds and clusters of buyers, shoppers, and just lookers. The palatial ground floor was dominated by a grand stairway and a great rotunda brightened by daylight which streamed through an overarching glass dome. Up and down these stairs, frequenting the high-ceilinged grandeur of these consumers' palaces, came the lords and ladies of these domains by the thousands and tens and hundreds of thousands.

The traditional elegance of the grand stairway was complemented by the modern charm of the elevator, which made the upper floors more easily accessible. Incidentally, the elevator car pushed together in sudden intimacy random members of the public who had the same destination. Elevators had been tried before for freight, and there had been experiments in using them for passengers in hotels. But the department store gave everybody a chance to enjoy them.

The essential problem was to combine speed and safety. The old freight elevators, in which the cage was counterweighted by a plunger that descended into the ground to a depth equal to the height of the building, was relatively safe but slow. To obtain faster movement it was necessary to use a system of pulleys, which increased the wear on the ropes holding up the cage. This increased the danger of a plummeting cage. Then, to insure against such accidents, Elisha Graves Otis, an ingenious New Englander who had been born and raised on his father's Vermont farm, invented a safety device. He set up ratchets along each side of the shaft and attached teeth to the sides of the cage. These teeth were held clear of the ratchets by the rope which held up the cage, but when the rope ceased to be in tension, the teeth were released against the sides of the shaft and gripped the cage safely in place. Otis himself sold the public on his device at the Crystal Palace Exposition in New York City in 1854. He had his elevator drawn up, then he melodramatically cut the supporting rope and displayed himself in the cage safely held in place.

It was in the Haughwout Department Store in 1857 that Otis first put his safety elevators into permanent use. Experiments with the elevator had been made in hotels as early as 1833, and the old Fifth Avenue Hotel in 1859 installed a practical passenger elevator. When Strawbridge & Clothier in Philadelphia carried its customers up and down in an elevator in 1865, anybody could enjoy free of charge this novel sensation. Otis patented a steam-powered elevator in 1861. By the time the Eiffel Tower was built for the Paris Exposition of 1889, three hydraulic elevators (one made by Otis) arranged in stages carried a visitor to the top in seven minutes. Even faster were the new electric elevators, which first appeared that year and which soon were carrying the public in Macy's and Wanamaker's.

Glass would play an important new role in this new consumer's world. Before the introduction of electric lighting, large windows were needed to bring daylight into the extensive buildings. But at least until the mid-nineteenth century, large sheets of glass were costly and difficult to make. "Plate glass" (the word came into English about 1727), a flat sheet smooth and regular enough for mirrors or large windows, was made from a rough sheet of glass which was then laboriously ground and polished. At first the rough sheets were produced by blowing (which could make a plate no bigger than about 50 inches by 30 inches); then, in the early eighteenth century, the French perfected a system of casting glass in sheets. In 1839 an Englishman simplified the process for removing irregularities. Further improvements pointed the way to the continuous plate-glass process using rollers, which could make sheets of any length with the transparency of the old plate.

The larger sheets of glass, combined with the light cast-iron frame of the building, transformed the ground floor of department stores. The windows at street level were no longer merely openings to admit light and sun, but vivid advertisements—literally "show windows," an Americanism which came into use about the mid-nineteenth century for a shopwindow in which goods were displayed. The shop itself, the stock, and the goods themselves had become a powerful new form of advertising. Now for the first time the society's full range of material treasure would be laid out for all to see. "Window-shopping" was the name for a new and democratic popular pastime. The effectiveness of a building, the desirability of a retailing location, were now measured by the numbers in the passing crowds.

These urban crowds were brought to the city center by two important devices, neither of them quite new, but both newly flourishing in the United States after the Civil War. One made it easier for people to come to the department store; the other stirred them with the latest merchandising news, arousing their desire to come.

Public transportation did not appear in American cities until the second half of the nineteenth century; until then the ordinary citizen commonly shopped within walking distance, that is to say, within a radius of about two miles. Except for wealthy customers who could afford their own carriages, or for visitors from afar, a city merchant drew his customers from those who could walk to his shop from their house. This helped explain the importance of the neighborhood community. Almost all a man's activities, including his buying and selling, were with people who lived nearby and who as neighbors were very likely known to him personally. A neighborhood community was a walking community: of passers-by, of casual streetcorner encounters, of sidewalk greetings and doorway conversations.

Streetcars in the cities helped change all this. The early alternatives to walking were the omnibus (a kind of city stagecoach which held few passengers, was expensive, appeared infrequently, and lumbered slowly over the streets of cobblestone or mud) or the steam-driven railroad. Although the railroad was speedy, the noise, smoke, and embers from the locomotive made it a menace on the streets, and it was not suited to a line with frequent stops. The first effective public transportation within cities was the horse railroad, whose level tracks made the ride more comfortable, and which was well adapted to stop at any corner. We have become so accustomed to public transportation in our cities that we forget what a revolution in city life came with the first cheap public transportation.

The revolution occurred in many places at about the same time. As good an example as any other is the story of Boston, which has been admirably told by Sam B. Warner, Jr. In 1850, congested urban Boston extended out only about two miles from City Hall. By 1872 the horse railroads had pushed the radius out another half-mile. By 1887 the horse-car had pushed on for still another mile and a half, doubling the 1850 radius, and incidentally, of course, quadrupling the area of dense settlement. When by the 1890's the horsecar was displaced by the electrified trolley, which moved twice as fast and could carry three times the number of passengers, public transportation reached out for at least another two miles, now making a greater Boston that reached six miles from City Hall.

The profits and enthusiasms of suburban investors and streetcar builders accelerated the process. The first street railway in Greater Boston, a single car in 1852 running between Harvard Square, Cambridge, and Union Square, Somerville, was so profitable that it invited imitation by other investors. It seemed simple enough to lay tracks on the roadbeds already provided by the city, to mount a coach on the rails, and buy a horse or two to provide the power. Booster real estate men who had bought tracts on the edge of the city had as much interest in linking their lots to the city centers as the earlier boosters of upstart towns had in bringing the railroad their way. Optimistic businessmen like Henry M. Whitney, the steamship magnate who consolidated the Boston lines in 1887, tried to attract more passengers by a standard five-cent fare and free transfers.

Meanwhile the boosters for streetcar monopolies urged the great "moral influence" of street railways. At long last, they said, the workingman who had been crowded into a multifamily tenement in the congested center of the city could buy his own lot, build his own house, and enjoy the wholesome delights of the rural suburb. The rapid expansion of street railways brought a scramble for franchises and entangled urban politics in the quest for monopolies, what Lincoln Steffens called *The Shame of the Cities* (1904). But regardless of the motives, the result was to draw more customers into the orbit of the city.

Streetcar tracks were rigid channels. A man in a streetcar had to go where it took him. And the streetcar, in almost any city, was likely to take him into the center; there were the great consumers' palaces.

Along with the centralizing influence of the streetcar, which brought city dwellers to department stores, came a new indrawing power over customers' minds and desires: the daily newspapers with large circulation concentrated in the cities. The department store, through its heavy newspaper advertising, contributed substantially to the success of these papers, and so helped keep them independent of subsidy by political parties. In this way the department store, like other large advertisers, indirectly contributed toward the political impartiality of American news reporting that would contrast sharply with the partisan-dominated press of France, Italy, and some other countries. The urban dailies also did much to help the great consumers' palaces to attract their vast constituencies. Just as the rise of the suburbs in the late nineteenth century was inseparable from the story of the streetcar, so the rise of the department store was one with the rise of newspaper advertising. The department-store pioneers were pioneers in the art and science of advertising.

R. H. Macy, like the mail-order pioneer Richard Warren Sears, was a bold and ingenious advertiser in the days before merchants had made advertising a part of their regular operations. Macy used repetition, composed bad verse, and combined hundreds of tiny agate-sized letters (the only kind which newspaper editors tolerated at the time) to make the Macy star or to produce larger letters. Beginning in 1858, he dared to leave large white spaces in the expensive columns; he advertised frequently, and put his ads in four or five different papers at the same time, to overshadow his more conservative competitors. John Wanamaker of Philadelphia was another vigorous leader. He pioneered in 1879 with his first full-page newspaper advertisement; within ten years Wanamaker's full-page advertisements were appearing regularly. Other department stores followed, and big-city dailies all over the country profited. In 1909, when Wanamaker's in New York City began putting full-page advertisements in the evening newspapers daily, this gave the lead to the evening over the morning newspapers in advertising linage. In Chicago, too, Marshall Field had become a big newspaper advertiser. Mandel Brothers made news when it contracted with the Chicago *Tribune* to run its full-page advertisement six days a week throughout 1902, for an annual fee of $100,000.

By the beginning of the twentieth century, the department store had become a mainstay of the big-city daily newspaper throughout the country. And as the circulation of dailies increased, the dailies became the mainstay of department stores, the increasingly powerful enticers of their hundreds of thousands of customers. City newspapers had become streetcars of the mind. They were putting the thoughts and desires of tens of thousands of people in the new cities on tracks, drawing them to centers where they joined the hasty fellowship of new consumption communities.

The department store, as Émile Zola observed in France, "democratized luxury." We have forgotten how revolutionary was the new principle of free admission for the whole public. In the old fairs and bazaars, the stall keepers had of course shown off their goods to the passing crowds. But the goods displayed to the common view were of the familiar sorts, to satisfy familiar wants. Any passer-by could look at the fruits and vegetables, at the sides of beef or the slabs of pork, at pots and kitchen utensils, at a basket or a length of cheap cloth. The costlier textiles or home furnishings were kept in an inner room, to be brought out only for serious customers who could afford such goods. In the great cities of the world, the better shops hung their symbols over the door, but they boasted their exclusiveness, displayed the coat of arms of the noble family who had appointed them to be their supplier, and exhibited little or none of the merchandise to the casual passer-by. The less expensive shops, too, were specialized, and their stocks of ready-made goods were small. In the latter part of the eighteenth century "shop" became a verb: then people began to "go shopping"—that is, go to the shops to see what they might buy. But still, common citizens might spend their lives without ever seeing a wide array of the fancy goods that they could not afford.

The department store helped change all this. Now a flowing, indiscriminate public wandered freely among attractive, open displays of goods of all kinds and qualities. One needed no longer be a "person of quality" to view goods of quality. Anyone could enter a department store, see and handle the most elegant furnishings. In this new democracy of consumers it was assumed that any man might be a buyer. Just as standard of living, by contrast with wealth, was a public and communal fact, so, too, buying and "shopping" became public. In the department store, as in the hotel, the distinction between private and public activities became blurred.

An urban shopper now could stroll through the world of actual goods as casually as a farmer soon would be leafing through the mail-order catalogue. Architects now aimed to make goods into their own advertisement: a permanent exposition for consumers and would-be consumers. Formerly merchandise had remained mostly dispersed into its raw forms, awaiting a customer's command or design. But this world of the ready-made was now a world of "consumers." Goods that had been assembled in advance into shoes, suits, or furniture were offered enticingly to the whole milling passing public. In these palaces of awakening desire, the new merchandisers hoped to offer something near enough to what the customer might already have wanted, and to stir him to wants he had never imagined.

In other, subtler ways, the market was homogenized and democratized. One of the most interesting, and least noticed, was the fixed-price, one-price policy of the great new department stores. The old practice, still a spice of life in the world's bazaars, was for the seller to bargain individually with each buyer, asking a price determined by that particular buyer's social position, his need, and his desire for that particular item. Some merchants marked each item with its cost (in private symbols), and then sought to secure from the customer the highest price above that which he could manage to extract. The refusal to bargain was considered churlish or unsociable, and it surely made life less interesting. The price actually paid for an item varied with the bargaining ability of each customer.

It is not surprising, then, that doctrinaire egalitarians had objected to this way of pricing. George Fox, founder of English Quakerism, as early as 1653 urged his followers to refuse to haggle, and advised merchants to fix the one fair price for every item and for all customers. Like some other Quaker principles, this was considered odd, but it had its business compensations. Customers who distrusted their own bargaining ability, Fox himself explained, would be reassured by the thought that "they might sende any childe and be as well used as themselves at any of these [Quaker] shopps."

The progress of the fixed-price policy had been slow, but department stores were quickly committed to it. The pioneering Paris department store Bon Marché had a fixed-price policy as early as 1852. For the large American department stores the policy was inevitable. In 1862, when Stewart's already had a staff of about two thousand, most of them on meager salaries and personally unknown to the store owner, it was not feasible to entrust bargaining to the individual salesman. A consequence, then, was the democratization, or at least the equalization, of prices. One price for everybody! Regardless of age, sex, wealth, poverty, or bargaining power. The price was marked for all to see. As the merchandise itself had become public and the intimate shop had been transformed into a palatial lobby where the best merchandise was open to vulgar eyes, so, too, the price was no secret.

Goods were priced for mass appeal, and department-store services were offered to the general public: free delivery, freedom to return or exchange goods, and charge accounts. These services, like "Satisfaction guaranteed or your money back" (an early department-store slogan), were not a product of private promises between shopkeeper and customer, but were part of a "policy," publicly proclaimed and advertised, from the firm to all consumers.

In a new sense now every sale and every purchase became a public act. The consumer was accepting an offer made, not only to him, but to anyone, usually in advertising. And advertising developed into characteristic commercial relationship of the new age. Now it was longer buyer and seller, the custom maker and the customer. It w advertiser and consumer: much of the advertiser's appeal was in his bigness; the consumer was a numerous horde whose strength was in numbers. The consumer now was being persuaded not merely to become a customer but to join a consumption community. He was being offered something that was not just for him but for everybody like him, and as both advertiser and consumer knew, there were millions.

NATIONWIDE CUSTOMERS

Just as department stores drew together thousands within the city in their consumers' palaces, other new enterprises reached out from city to city, creating nationwide consumption communities. Chain stores, pioneers of the everywhere community, built communities of consumers across the land. The expression "chain store," an Americanism firmly settled into the language by the beginning of the twentieth century, described one of a group of similar stores under common ownership. This was not, of course, a new idea, nor an American invention. But in the United States in the century after the Civil War, the chain store became a newly powerful institution.

"Cash-and-carry," an Americanism added to the language by the early twentieth century, would become the motto of the chain stores. An affirmative way of saying "no credit and no deliveries," it would be an advertising slogan to inform all would-be customers that here was a shop with no frills, where the customer could save money. The department stores, oddly enough, had succeeded by offering some of the personal conveniences traditionally associated with the small neighborhood shop and the friendly reliable shopkeeper. Their developing systems of credit and installment buying and their numerous incidental services would actually provide a foil for the sales pitch of the new chain stores, which usually featured price and made a public virtue of their economies.

The first unit of what by mid-twentieth century was to be the chain-store system with the largest annual volume was founded in 1859. In that year George F. Gilman and George Huntington Hartford, both from Maine, opened a small store on Vesey Street in New York City under the name of The Great American Tea Company. By cutting out middlemen, by buying tea in quantity and by importing it themselves from China and Japan, they offered tea at the spectacularly low price of 30 cents a pound when others were charging $1. They attracted customers by Barnumesque showmanship: premiums for lucky customers, cashiers' cages in the shape of Chinese pagodas, a green parrot in the center of the main floor, and band music on Saturdays. They sent eight dapple-gray horses pulling a great red wagon through the city and offered $20,000 to anyone who could guess the combined weight of the wagon and team. They gradually added other grocery goods—spices, coffee, soap, condensed milk, baking powder—and by 1876 had multiplied their stores to the number of sixty-seven.

With a booster enthusiasm worthy of an upstart Western town, they anticipated greatness by adopting the name The Great Atlantic & Pacific Tea Company in 1869. Perhaps the notion was that this chain of stores would unite the two oceans as did the Union Pacific Railroad, which had been completed that same year. The number of stores increased and the distinctive red-and-gold façade became familiar across the land. By 1912 there were nearly five hundred A & P stores. While they offered the advantages of lower prices which they claimed came from large-scale purchasing and from the elimination of middlemen, they still provided charge accounts and free delivery.

Led by John Hartford, son of the founder, the great expansion of the A & P chain came in 1912. Between 1912 and 1915, a new A & P store was opened every three days, to a national total of one thousand stores. Expansion was based on the cash-and-carry idea and on reduction of staff to make the one-man "economy" store. Meat, which soon became the largest single item, was not added till 1925. For the year 1929, total A & P sales exceeded $1 billion; in the following year, A & P stores numbered 15,709. By 1933, A & P was doing over 11 percent of the nation's food business. After that year, there was a trend to larger stores, and the number of individual stores gradually decreased. But by 1971 the 4,358 A & P stores reached an unprecedented annual sales volume of nearly $5.5 billion.

The builders of these new nationwide consumption communities met bitter opposition from local merchants, hometown boosters, and champions of neighborhoods who stood for the *local* community. The keepers of the old general stores had fought the big-city department stores; they would also fight RFD, they opposed parcel post, and they attacked the mail-order "monopolies." The menace of "chain stores," they said, was a threat to the whole American way of life. By the early 1920's, when a number of chains were prospering, the opposition of small, independent retailers became organized. The National Association of Retail Grocers in their annual convention in 1922 urged laws to limit the number of chain stores in any community. In defense of the neighborhood store they proposed various legal devices, such as special escalating taxes on every store beyond the first under the same ownership in a given state, and special taxes on merchandise purchased by chain stores.

At one time or another most states enacted some type of anti-chain-store tax. The most extensive effort at controlling chain-store merchandising was the Robinson-Patman Act of 1936 (sometimes called the Federal Anti-Price Discrimination Act), a New Deal measure amending earlier antitrust legislation. It aimed at the chain-store practices of "price discrimination" which were said to destroy competition or promote monopoly, and it gave the Federal Trade Commission important, if vague, supervisory power. The chain stores, like other large enterprises, had been guilty of some abuses. But they were unstoppable institutions in the movement to larger and larger consumption communities. The anti-chain-store movement, like the anti-RFD and anti-parcel-post movements, was a rearguard action. Its spokesmen spoke for the dying past of the general store, the village post office, the one-room schoolhouse and the friendly corner drugstore.

Jeremiads against the chain store really expressed bewilderment at the dissolving of the neighborhood community. "The chain stores are undermining the foundation of our entire local happiness and prosperity," lamented the Speaker of the Indiana House of Representatives in a letter to his constituents in the late 1920's. "They have destroyed our home markets and merchants, paying a minimum to our local enterprises, sapping the life-blood of prosperous communities and leaving about as much in return as a traveling band of gypsies." Senator Royal S. Copeland of New York declared, "When a chain enters a city block, ten other stores close up. In smaller cities and towns, the chain store contributes nothing to the community. Chain stores are parasites. I think they undermine the foundations of the country."

A hysteria which paid heavy political dividends seized the congressional representatives from the rural and small-town world. "A wild craze for efficiency in production, sale, and distribution has swept over the land," warned Senator Hugo L. Black of Alabama in 1930, "increasing the number of unemployed, building up a caste system, dangerous to any government. . . . Chain groceries, chain dry-goods stores, chain clothing stores, here today and merged tomorrow—grow in size and power. . . . The local man and merchant is passing and his community loses his contribution to local affairs as an independent thinker and executive."

The response of the chain stores was multiplex. Their owners tried to answer the accusations that they lacked old-fashioned community loyalty by going to great lengths to advertise locally and to reward examples of community leadership among their managers. In 1939, under the very shadow of the Robinson-Patman Act, the trade journal *Chain-Store Age* announced awards for "Community Builder of the Year" to advertise the hometown services of chain-store managers. The local manager, they argued, actually did support the Community Chest and the Red Cross, he helped local students, he served his local church, and he cooperated with local merchants.

But both the accusation and the response were beside the point; the chain store announced and symbolized a new kind of community.

The new consumption communities were, of course, shallower in their loyalties, more superficial in their services. But they were ubiquitous, somehow touching the American consumer at every waking moment and even while he slept. Senator Black was right in his alarums that the "local man" was passing. Man was no longer local. As the American population adopted mobility as normal, the new arrivals in a new suburb or city who might not know their neighbors would at least feel somewhat at home in their A & P (where they knew where to find each item) or in their Walgreen's (where familiar brands abounded). Had these enlarging communities of consumers provided some slight solace and substitute for the declining neighborhood community?

HOMOGENIZING SPACE

By the mid-twentieth century, millions of Americans were living and working high in the air. They spent whole days and nights in towers of steel and glass, and the skyscraper became a symbol of the American city. Just as nineteenth-century boosters had boasted of their hotels, their Palaces of the Public, so twentieth-century boosters boasted of their towers. Both were as much a product of hope and aspiration as of necessity. The towering skyscraper expressed a new, latter-day American boosterism, a determination to compete with Nature herself, to win over the limitations of matter and space and seasons.

The grandest went up in the largest, most congested cities. New York City's Woolworth Building (1913) and her Empire State Building (1931), and Chicago's Tribune Tower (1925), and many others responded to the real or imagined needs of the great metropolis. But in smaller cities, too, all over the nation skyscrapers shot up. For the American skyscraper was not simply a reflex response to economic need or an answer to the scarcity of land. In Tulsa, Oklahoma, for example, and in other cities of the West, mini-skyscrapers rose in the midst of endless acres of uninhabited prairie.

As Old World cities grew they had spread out. The most populous cities were inevitably the most extensive. Rome was spread over seven hills, and before the middle of the twentieth century, Greater London covered nearly 700 square miles. In European cities the tallest buildings, except for monuments, exposition towers, and occasional tours de force like the Eiffel Tower, reached up only five or six stories.

The tall structures that Americans built were not mere Eiffel Towers. They were buildings to work in and live in, for Americans had developed new ways of indoor living, ways of homogenizing space and the seasons. Americans could cease to be earthbound because they were no longer bound to the earth for their drinking water and washing water, for disposing of human wastes, for their means of keeping warm and cool, for their ways to communicate with neighbors. The skyscraper was the climactic symbol of man's ability to rise above particular places and times to satisfy his needs, to keep himself comfortable and at work, making experience for all Americans, wherever they lived, more alike.

The first large American community water supplies were not motivated by the desire for household convenience, but rather for a variety of public purposes—more and better drinking water, water to flush the streets and to fight fires. The first sizable municipal waterworks in the United States, which brought the waters of the Schuylkill to Philadelphia in 1801, was a response to the city's recent yellow-fever epidemics that had decimated the population, forcing thousands to flee to the countryside. It was believed that disease could be prevented by flushing the streets daily, and especially in hot weather. The designer of the waterworks, the eminent Benjamin H. Latrobe (whom Jefferson appointed as the first surveyor of United States public buildings, and who was to build the first section of the Capitol), had encountered an assortment of fears and prejudices. Latrobe's plan for the waterworks required the use of steam engines at the pumping stations in the heart of the city. Philadelphians, who had heard about the explosions of these new-fangled engines on Western river boats, imagined that the pumping stations would bring the perils of boiling water and steam into their own neighborhoods. Latrobe finally carried the day with his declaration that "a steam engine is, at present, as tame and innocent as a clock," and his scheme was executed at a cost of about a quarter of a million dollars.

But a full year afterward Philadelphia, which with its population of 70,000 was then the largest city in the United States, had received an annual revenue of only $537 from a total of 154 water takers. A decade later, when the population of Philadelphia had reached 90,000, the number of water takers had risen to 2,127. The idea that running water was a household necessity was at least a half-century in the future. City dwellers could still gather what water they needed free of charge from pumps or running streams. "It will be some time," one Philadelphian predicted, "before the citizens will be reconciled to *buy* their water." Paying for water to drink appeared almost as absurd as paying for air to breathe. But even that fantastic necessity seemed in the offing by the late twentieth century.

The earliest American community water supplies were brought by wooden conduits to a town or to a group of houses. During the first half of the nineteenth century such water suppliers were usually private companies authorized by state legislatures and municipal governing bodies and run for profit. Within three decades after the opening of the Philadelphia municipal waterworks in 1801, New Orleans, Pittsburgh, Richmond, and St. Louis took up the idea.

In New York it took failure by the private Manhattan Company, dominated by Aaron Burr, and a half-century of wrangling to bring the city round to providing a public water supply. Cholera epidemics in 1832 and 1834 and a scourge of fires in 1834 estimated to cost the city $1.5 million pushed the city into action. "We are at present," a New Yorker observed, "supplied with spring water carted round in carts and brought from the upper parts and suburbs of the city. This water, although far from good, is much better than that obtained from wells in the city; for this we pay at the rate of two cents per pail, three pails per day is but a moderate quantity for a family, and three pails per day costs twenty dollars per annum." City dwellers commonly spiked their unpalatable water with spirits and incidentally added the temperance argument to all the others. "Water is one of the elements," a New Yorker found it necessary to argue, "full as necessary to existence as light and air, and its supply, therefore, ought never to be made a subject of trade or speculation." The fire of December 17, 1835, the worst in the city's history (inhabitants compared it to the conflagration of Moscow) dramatized the need.

The great Croton Aqueduct brought water thirty miles south from a reservoir dammed up from the Croton River, and on October 14, 1842, began pouring millions of gallons into New York City. "Nothing is talked of or thought of in New York but Croton water," Philip Hone noted in his diary; "fountains, acqueducts, hydrants, and hose attract our attention and impede our progress through the streets. Political spouting has given place to water spouts, and the free current of water has diverted the attention of the people from the vexed questions of the confused state of the national currency." But some New Yorkers, complaining that the water was "all full of tadpoles and animalculae," were "in dreadful apprehensions of breeding bullfrogs inwardly."

The most powerful argument for improved municipal water supplies came from London, when an English physician, John Snow, showed that the victims of that city's cholera epidemic of 1849 had all drunk water from a certain Broad Street pump. Snow proved that bad water was not simply water that did not sparkle and had an unpleasant odor, but it contained specific agents which came from feces. Even after proof that specific diseases could be traced to specific impurities in water, purifying systems came only slowly to American cities.

The notable American contribution was made by James P. Kirkwood, who had been sent to Europe by the city of St. Louis to study the purification of river water. After reading Kirkwood's report, St. Louis city officials decided not to try to filter the muddy Mississippi. But the city of Poughkeepsie then commissioned Kirkwood to build a plant to filter waters of the Hudson. It was the first large-scale water filtration plant in the United States, and became a model for the world. One of the main problems in sand-filtration plants was the accumulation of scum and impurities which were difficult to remove from the sand-filtration beds. Kirkwood's ingeniously simple proposal was that instead of scraping off the impurities from the sand, the flow of water be reversed through the filter to backwash the impurities.

In one American city after another, water began to flow in seemingly endless streams. By mid-century, running water was commonly found in upper-middle-class households in large cities. In 1860, every one of the nation's sixteen largest cities (each with a population of at least 50,000) had some sort of waterworks. All but four (New Orleans, Buffalo, San Francisco, Providence) of these were municipally owned, and many smaller cities also commonly had waterworks.

Even while running water in the household was still an upper-middle-class luxury, those Americans who could afford it were already treating the water supply as if it were inexhaustible. Planners had made the mistake of basing their predictions on the early Philadelphia experience. The Bostonians who argued against the expense of the Cochituate Aqueduct (completed in 1848) had said that it was ridiculous to imagine that the city would ever consume 7.5 million gallons a day. Yet within five years, the city's average daily consumption exceeded 8.5 million gallons. What ran up the consumption was not that there were too many families connected to the system, but that, as the Boston Water Board estimated, two thirds of the water was being wasted. In livery stables, in primitive water closets and urinals, water was left running constantly. In cold weather, when householders feared that their pipes might freeze, they kept their faucets wide open all night. During a cold spell in January 1854, daily consumption rose to 14 million gallons, the reservoirs were nearly dry, and houses in the higher parts of the city were left without water. Inspectors were sent about the city at night listening for the sound of running water so they could warn offenders that their supply might be cut off. In 1860 the Water Board complained that Boston's average of 97 gallons daily for each inhabitant was "without parallel in the civilized world," and expressed fear that the city soon would be unable to supply the citizens' demands.

These municipal waterworks, like other new resources of flow technology, were not merely new means to supply one of man's ancient necessities. They became themselves channels which brought into being inexhaustible new demands. Even before all Americans had become accustomed to the luxury of running water, the municipal waterworks had become sources of new scarcities. The story of running water would, of course, be paralleled a half-century later by the story of running electricity.

Running water made possible a host of new uses for water, and the spread of many older uses to more and more Americans. The New York water commissioners warned New Yorkers only two years after the Croton Aqueduct went into operation that they had not intended to supply fountains in all the city's parks, nor to amuse all the city's boys with water flowing from fire hydrants. At first it was generally assumed that baths would continue to be primarily a public facility. In 1849, when Philadelphia already had 15,000 houses with running water, only about 3,500 had private baths. In New York the price of general admission to a public bath was three cents, or six cents if you wanted to bathe privately in a separate room.

In other parts of the world, too, a household bathroom was still a luxury. But in the United States within a half-century it would begin to be a middle-class necessity. The nation which was world headquarters for the democratization of comfort was naturally enough the home of the democratized bath. The progress of the American bathroom was rapid and spectacular. By 1922 Sinclair Lewis' Babbitt started every day in his "altogether royal bathroom of porcelain and glazed tile and metal sleek as silver."

In the older world, the *public* facilities tended to copy the *private*. Inns were shaped like large private residences, town halls were fashioned after the palatial dwellings of rich citizens. But the urban communities which sprang up in the United States in the nineteenth century were bristling with newcomers, while there were still few rich men and, of course, no ancient palaces. Here public buildings and public facilities made their own style, which gradually influenced the way everyone lived. The large display windows to show off the wares of new-style department stores eventually were adopted as "picture windows" in apartment buildings and in private households. The history of running water and bathrooms followed a similar pattern.

The luxurious American hotels, the Palaces of the Public, were among the first and most influential American buildings to bring running water indoors. Even before Boston had installed a municipal waterworks, the Tremont House, completed in 1829, had its own water system to feed its bathtubs and its battery of eight water closets on the ground floor. The rise of running water in America was literal as well as figurative. From the ground floor it rose gradually to the upper stories, where at first there were common bathroom facilities for the residents of each floor, and then it trickled out to each room. Generally, running water was brought first to the kitchen sink, then to the wash basin, and finally to the bathtub. But as late as 1869, Catherine Beecher and Harriet Beecher Stowe's *American Woman's Home,* a popular guide to home planning, showed a kitchen sink with water drawn from a hand pump.

In the United States, hotels continued to set the pace. As early as 1853, the luxurious Mount Vernon Hotel at Cape May, New Jersey, impressed Americans and amazed travelers from Britain by equipping every room not only with running water but also with a bathtub. By

1877, one medium-priced Boston hotel offered in each of its rooms a wash basin with running hot and cold water. But it was the early twentieth century before the private bathroom became normal for every room in better American hotels. When the enterprising Ellsworth M. Statler built his new hotel in Buffalo in 1908, his advertising slogan was "A room with a bath for a dollar and a half."

Running water was, first of all, a labor-saving device. In earlier times, even if there was a pump indoors at the kitchen sink, it still required muscle power to make the water flow; the wash basin or the bathtub required water to be carried by hand. When English travelers first saw plumbing fixtures in American hotels and households, they described them as simply another American response to the scarcity of menial labor. And the high cost of labor in the United States did help explain why running water and indoor plumbing so early became commonplace here.

Before there were running water and municipal sewers, the wash basin or the bathtub was moved about to wherever it was needed or wherever at the time the water could be most conveniently carried to them. There was no such thing as a "bathroom," since there was no functional reason why the bathtub and washstand should be kept in one place.

Water systems had begun as private enterprises, selling people the water they needed, but sewage-disposal systems could not start that way. This helps explain why they were so slow in coming. Running water could not become a common convenience until there were municipal systems of sewage disposal. In 1850 Lemuel Shattuck, the versatile New Englander whom we have already met as De Bow's collaborator in the census of 1850 and as a founder of American vital statistics, explained the urgent need for city sewage-disposal systems in his pioneer *Report of the Sanitary Commission of Massachusetts*. In 1860, when there were 136 city water systems in the United States, there were still only ten municipal systems for disposing of sewage. By 1880 there were more than two hundred. But at that time a sewage "system" was simply an arrangement for collecting and pouring unprocessed city sewage into nearby lakes or streams. The growth of city water systems in turn increased the urgency of the need for better systems of sewage disposal, since cities now commonly found themselves drawing their water supplies from the very streams they were polluting. Massachusetts kept the lead, by setting up a State Board of Health in 1869, and by pioneering in studies of sewage disposal at the Lawrence Experiment Station. But this was considered a nasty subject, embarrassing to discuss in mixed company. And it was only against considerable public resistance that the perils of human sewage, however diluted, for community water supplies were gradually demonstrated. At the Lawrence Station, Hiram

F. Mills developed a system of slow sand filtration for sewage, contradicting the popular notions of the day that rapid flow was what purified streams, and he thus helped reduce by 80 percent the Massachusetts death rate from typhoid.

As sewage systems matched water systems, there grew up a large new market for plumbing fixtures. The "room with a private bath" in American hotels and motels, and the multiplying baths in households, required better bathtubs, easier to clean and maintain. The first two decades of the twentieth century saw a sudden increase in the production of American enamelware. In the previous decades many materials had been tried: wooden boxes lined with sheet lead or zinc or copper; cast iron, plain, painted or galvanized; sheet steel; sheet copper; porcelain crockery; and even aluminum. But all these were either too fragile, too cold to the touch, or too expensive. A durable cast-iron enameled tub had been made as early as 1870, when a manufacturer was turning out one tub per day. Until 1900, sanitary fixtures were still hand-fashioned, but production had risen to ten bathtubs per worker per day. A few cast-iron enameled tubs were used in private Pullmans before 1900. Then by 1920, the one-piece, double-shelled enameled tub, destined to remain standard for a half-century, was being machine-made and mass-produced. The castings were serially poured, cooled, and scoured, and little skill was required of the workers. Between 1915 and 1921 the annual production of enameled sanitary fixtures (wash basins, bathtubs, etc.) doubled, and then doubled again by 1925, to reach over five million pieces.

The other essential element in the American bathroom, apart from the wash basin, was of course the water closet, or flush toilet. One reason for the slow introduction of indoor water closets for the disposal of human waste was that unlike the running-water wash basin or bathtub, it was not so obviously a labor-saving device. Not until near the end of the nineteenth century was it widely believed that human waste was a dangerous source of disease infection and water pollution. Rural life and mores discouraged the making of machines to do what nature could do better. The earth was supposed to have remarkable absorbing and deodorizing powers, and the common way of disposing of human waste was on the surface of the ground. People retired to a nearby woods or to some other sheltered place; and they tried to prevent *odors*, which were generally thought to be the real source of infection, by carefully selecting a different place each time. In villages it was found desirable to establish special private places or "privies." But privies were generally considered not so much sanitary devices as places of modest and dignified retirement. Even where there were privies, they were for use mainly by women and children, while the men still used the stable or the woods.

The first United States patents for water closets were issued in the 1830's, but it was past mid-century, after the rise of water systems and sewage systems, that the water closet began to come into general use. In 1851, when President Millard Fillmore reputedly installed the first permanent bath and water closet in the White House, he was criticized for doing something that was "both unsanitary and undemocratic." Even after there were city sewage systems, it was hard to persuade owners to go to the expense of installing plumbing and making sewer connections. Especially in the congested center of cities, where the problem of waste disposal was most pressing, it was often difficult to persuade landlords to use the sewage system. On sewered streets, thousands of persons were still using old privies and cesspools at the end of the nineteenth century.

Designing a satisfactory water closet was no easy matter, and it challenged the talents of plumbers, inventors, and hydraulic engineers. The problem was to design a bowl and apparatus that would be self-cleaning and not too noisy, and yet would not use excessive amounts of water. The modern wash-down closet that worked on the syphon principle, using suction to clean the bowl, had been invented in England by 1870 and was used by Americans who could afford it by the 1880's and 90's. The flush-valve toilet, which required less time between uses, was developed in the early twentieth century and long remained the only major improvement in toilet design. The water closet, like the bathtub, could be democratized only after the perfection of ways of mass-producing enamelware—plumbing fixtures of iron coated with enamel. By the opening of the twentieth century, American enamelware mass-produced for bathrooms had become good enough for European royalty, and the American product was installed in Buckingham Palace and in the private apartments of the king of Prussia. During the next decades, enamelware was displaced by porcelain.

In the United States a by-product of the widespread installation of water closets which required gallons of water at each flush, and another symptom of the democratization of everything, was the enormous increase in the public demand for water. As early as 1860, when it was found in Boston that the Parker House was using over 20,000 gallons a day and the Tremont House over 25,000 gallons, water meters were installed so that users could be charged for the actual amounts they consumed. Throughout the next century, novel household uses of water multiplied, with the washing machines, dishwashers, garbage-disposal machines, automatic lawn sprinklers, humidifiers, and air conditioners. Even before World War II, when the average per capita use of water in ten large European cities (including London, Paris, and Berlin) was 39 gallons per day, in ten large American cities (including New York, Philadelphia, Baltimore, Chicago, and Detroit), the per capita water use was four times as much, or 155 gallons per day. The new ways to use and to waste water seemed endless.

The new communal sources of water, and the communal outlets for sewage, became the unexpected causes for isolating individuals, incidentally changing social attitudes toward bodily functions. The old-fashioned privies, even in castles, were often designed so that their users could enjoy the company and conversation of fellow-users. The early American outhouses, too, commonly had more than one seat, to facilitate use by more than one person at a time. But the indoor toilet, partly because of its cost, was a loner. While in England, as Giedion has explained, the earliest bathrooms were large rooms, as spacious as the others in the house, in the United States the bathroom became a cell, distinguished by its compactness and insulation. This was a far cry from the bathing facilities of ancient Rome or Greece or modern Islam or Japan, where the common water source brought people together. And a far cry, too, from the sociability and gossip of the women at the well which had been proverbial ever since Rebecca's day. For Americans, the democratization of the bath meant the *private* bath.

As these mechanical devices using communal resources—water, gas, and electricity—became common in American households, they became the central unit around which buildings were planned. And as the American housewife went from the coal stove to the control panels, the engineer began to take precedence over the architect.

The skyscraper apartment or office building could hardly have become a year-round dwelling without central heating. For the problems of constructing a chimney to each of hundreds of rooms and then hoisting fuel for individual fires might have made large, high buildings impossible. And in the United States central heating, too, spread from public buildings into private households: another by-product of a fluid new world, where facilities open to the public aimed to democratize luxury. The Palaces of the Public, the luxurious American hotels, had introduced Americans to central heating before the mid-nineteenth century. The early heating plants were hot-air systems fired by wood or coal, but before the end of the century the boilers used in steam engines were being adapted for heating. And the exhaust steam from factory engines was piped into systems that warmed factories, offices, and meeting rooms. The Chicago public schools installed a steam-heating system in 1870.

The "radiator" was an American development. And the use of the word for a fixture through which steam or hot water from a central-heating plant circulates in order to heat a room is an Americanism dating from this era. In 1874 an American, William Baldwin, devised an improved radiator consisting of short lengths of one-inch pipe screwed into a cast-iron base. The widespread introduction of central heating into American households did not come until after the mass production of cast-iron radiators was developed in the 1890's. The American Society of Heating and Ventilating Engineers, founded in 1895, built laboratories at the land-grant University of Illinois, where pioneer experiments set the pace for the nation. By 1950, some sort of central-heating plant was found in half the nation's homes, a 50 percent increase within the preceding decade. Home-heating plants were increasingly fired by fuel flowing from some central source: along with oil, the use of gas or electricity was skyrocketing. By 1960 most American homes were heated neither by coal nor by wood. New systems, including infrared radiant heating and heating by atomic power, were being tried. The latest step in the blurring of boundaries between indoors and outdoors was the increasing use of underpavement pipes to keep streets and sidewalks clear and dry in winter.

Some kind of heating was a simple necessity for living and working in most of the nation. At first air conditioning (cooling, humidifying, or dehumidifying) was a mere convenience. And just as it was far more difficult to devise a machine to keep food cold than to cook, so, too, it was far easier to devise systems for warming rooms than for cooling them. But by the 1970's new styles of architecture with fixed windows and large panes of glass were beginning to take air conditioning for granted. Even before the Civil War, as we have seen, some American efforts had been made to devise a working system of artificial refrigeration. But these were mostly aimed at cooling the air of rooms to cure and prevent fevers.

An air-conditioning system feasible for general use would be a byproduct of efforts to solve certain specific problems of industry. Textile manufacturers had found that to keep their fibers soft and stretchable, to prevent the broken ends that required costly stoppage of their machines, they had to control the moisture content of the yarns. A fiber with just the right percentage of moisture was strong and pliable, and to dampen the fiber properly was called "yarn conditioning." In 1906, when an American textile engineer invented a system for accomplishing this by controlling the humidity in the air, he called it "air conditioning," and the name stuck.

For air conditioning, unlike many other innovations in indoor living, the theoretical as well as the practical advances were made in the United States. The man who developed the theory, who devised the machinery, and then envisioned the human possibilities of air conditioning, was Willis H. Carrier. In 1902, just a few months after Carrier received his engineering degree from Cornell, he noticed that the plant in Brooklyn where the humorous magazine *Judge* was printed could not line up their paper properly during the summer. The magazine covers were printed in color, and as the humidity varied during the summer, the paper stretched and shrank so that the successive colors applied by the machines did not "register" precisely one on top of the other.

With a flash of insight and ingenuity that would make him the founder of a great new industry, Carrier focused on the crucial relations between temperature and humidity. And he devised a machine to control both temperature and humidity to solve the plant's problem. In his "Rational Psychrometric Formulae; Their Relation to the Problems of Meteorology and of Air Conditioning," a paper he read to the American Society of Mechanical Engineers in 1911, he provided the theoretical basis for twentieth-century air conditioning. He then developed the devices needed to make air conditioning feasible in tall buildings and in individual residences. His new centrifugal compressor for refrigeration in 1923 reduced the number of moving parts, cut down noise and maintenance costs, provided a more compact machine, and finally extended the reach of air conditioning to the tallest skyscrapers. But as buildings grew larger, new problems appeared: it was impossible to provide enough ceiling space between the floors to accommodate all the ducts. Carrier solved this problem, too, by an ingenious new system of small high-velocity ducts which sped the air to room units which cooled or heated the air as required and diffused it around the room. For feasible home units the next requirement was a safe, noninflammable refrigerant, which the DuPont Company developed in 1931 in the form of "Freon 12."

Chocolate factories, which formerly had to close down during hot, humid periods, now, when air-conditioned, could operate regardless of the weather. Bakeries and chewing-gum factories, tobacco factories and ceramic factories, printing plants and munitions factories—all of which had been menaced by changes in the weather—were now free to make their own weather and so keep their production lines moving. The New York Stock Exchange installed a rudimentary air-conditioning system in 1904, followed three years later by New York's Metropolitan Museum of Art. A theater in Montgomery, Alabama, installed a system in 1917. In 1922, Grauman's Chinese Theatre in Hollywood installed a new-style system by which the conditioned air was poured in at the ceiling and the stale air was removed in ducts under the seats. Air-conditioned railroad cars came next. The comforts of the movie theater in the 1930's awakened Americans to their "need" for air conditioning in their offices, factories, shops, and homes.

Once again machines and techniques were spreading from the public commercial world into the private personal world. Room air conditioners, produced in small quantities in the 1930's, had become big business by the 1950's. In the 1960's more than three million such units were being produced annually, three-fourths for use in homes. By the mid-twentieth century, central residential air conditioning had become the industry's biggest market, and before 1970, $1 billion worth of central residential air conditioning was being installed every year. Even in automobiles, air conditioning was beginning to become standard equipment.

"Every day a *good* day" had been Carrier's motto in 1919 when he advertised that a visit to an air-conditioned movie theater "imparts the same splendid physical Exhilaration you would feel after a two-hour vacation in the naturally pure air of the Mountains." He described hotel dining rooms "where even the Air is Appetizing." And he predicted that "*you* will see the day when Manufactured Weather is making 'Every day a *good* day' all over the land, in every type of building from the modest bungalow to the spreading industrial plant enclosing millions of square feet of space. Mark my words."

Carrier's booster optimism was more than justified. Americans, in the 1960's and 70's, became accustomed to air conditioning in their schools and working places and homes, and they began to expect it wherever they went. Vast shopping centers, like Rochester's Midtown Plaza, Savannah's Oglethorpe Plaza and Dallas' North Park Mall, each of which enclosed nearly 1.5 million square feet, were air-conditioned. Imaginative planners like Walt Disney and his colleagues explored the possibility of city centers entirely enclosed and air-conditioned.

A climax of the new stage of indoor-outdoor confusion came appropriately enough in 1965 with the building of the Houston Sports Astrodome which aimed to bring an entire 650-foot span, large enough to enclose a baseball diamond or a football field, under air conditioning. For this purpose there was an elaborate set of testing instruments: outdoors, a pyroheliometer on the roof to measure the angle and intensity of the sunlight, and an anemometer to record the velocity and direction of the wind; indoors, an ultraviolet sensor to register the density of smoke and dust in the air, and to note variations of visibility, humidity, and temperature in all parts of the arena. The clear plastic shell, admitting sunlight to the ground inside, actually allowed real grass to grow on the playing field, but this produced a blinding glare for the baseball player who looked up to catch a high fly. When the plastic was painted over to keep out the sun glare, the grass died. To preserve an outdoor effect, this stubble was at first painted green. Then plastics technicians devised a new kind of artificial grass, "Astroturf," which had the appearance of grass from a distance, and for playing purposes had resilience and the other qualities required. For sports purposes, Astroturf soon proved in many ways superior to the real thing, and was then installed on some outdoor football fields and outdoor "lawn" tennis courts; it even began to appear in some city lawns.

There was less difference than ever before between what man could do indoors and what he had to go outdoors and brave the weather for.

There were all sorts of new indoor ways of getting water to drink, of disposing of human waste, of enjoying sports. Americans began to carry their indoors with them when they listened to stereo music and radio news in their air-conditioned auto capsule or on the beach. The common-sense distinction between outdoors and in, between the world Nature's God had made and man's little artificial world was blurred as never before, leaving Americans more disoriented than they commonly realized.

CHAPTER 2

The Resisted Revolution: Urban America and the Industrialization of Agriculture, 1900–1930

David B. Danbom

America's urbanization in the early decades of the twentieth century, the critical period in which the outlines of modern America were formed, has interested historians far more than the history of rural change in the same period. An exception is David B. Danbom, whose *The Resisted Revolution* performs an important service in recovering an overlooked chapter in the modern history of technology and society. Danbom's book narrates the story of urban reformers' attempts to make agriculture scientific and efficient, and to transform the family farm into a business.

The first portion selected, "Action, Inaction, and Reaction," concerns the period 1914–20, when the Smith–Lever Act (1914) provided federal funding for a network of government "extension" agents who were to take the latest scientific and technical knowledge to farmers and farm housewives. Federal funding for the extension agents was, by design, only partial; the counties in which the agents were to serve were asked to provide the remaining part of the necessary funds, and government planners thought this arrangement would ensure that the host communities would have a vested interest in being receptive to the extension agents' advice, since they were paying a part of these county agents' salaries.

Farm communities did not, however, prove receptive to the extension agents. Danbom shows that farmers were distrustful of scientific agriculture and its "missionaries," and for reasons that Danbom finds reasonable and rational. For example, the farmers knew that their traditional methods resulted in crops with modest yields, but the methods were reliable and carried little risk of failure. The new methods touted by the extension agents brought the promise of increased yields but also brought a much higher risk of crop failure due to the farmers' unfamiliarity with the new methods or to the nature of the new crop variety itself.

Farmers, long known for and often proud of their conservatism, were skeptical of change in general, but they were especially skeptical of change introduced by these unfamiliar outsiders, who were sent by the Country Life Movement, an urban-based group concerned about rural inefficiency and poverty, and by the USDA (U.S. Department of Agriculture). The farmers knew that the county agents' primary goal was to increase agricultural production, which was to benefit primarily urban consumers, not farmers. The county agents also tried to organize farmers, but when organizers ignored existing rural organizations, they found that it was extremely difficult to find either willing leaders or followers.

Acceptance of technological change does not advance smoothly or automatically, and Danbom shows us how a nontechnical factor, rural distrust of the cities, played a crucial role in slowing the dissemination of technical innovations. But by 1930, the industrialization of agriculture had proceeded despite farmers' resistance. In the selection entitled, "The Price of Progress," taken from the final chapter of Danbom's book, the author assesses the hidden values contained in the new technology and the related call for agricultural efficiency.

Farmers were asked not simply to adopt new techniques; they also were asked to change fundamentally the ways in which they lived and regarded their work. They were asked to turn into businessmen who needed special expertise and financial capital to survive. Thus entry was blocked to the one occupation that had traditionally been open to the ordinary person who lacked skills and capital.

Danbom makes no secret of his contempt for the bureaucrats who were responsible for the industrialization of agriculture, the technocrats who were solely concerned with efficiency and heedless of the negative impact of their programs on rural life and values. One of the oldest beliefs in this country is the agrarian myth, the belief that rural life, by its simple rhythms, its close physical relationship to nature, and the premium placed on an individual's own hard work, creates higher moral standards than found in cities. In the 1920s, as the small family farm began to be displaced by large-scale, businesslike operations, the concern was raised that America was losing an important repository of goodness and morality. Danbom takes this concern seriously.

Today, less than three percent of the American work force remains in the agricultural sector, as the industrialization of agriculture seen in the early twentieth century has continued. But the issues discussed in Danbom's conclusion have a contemporary ring, for we continue to

debate whether the government should continue farm subsidy programs, which are intended partly to help preserve the remnants of small-
scale family farming and the special values it allegedly preserves intact.

ACTION, INACTION, AND REACTION

• • •   T he cast of industrial missionaries to rural America was
completed when the Smith–Lever Act institutionalized
the county farm and home demonstration agent system. Supporters of
extension envisioned these men and women agents reorganizing rural
society, instructing rural children in the ways of advanced agriculture,
and making scientific farmers and homemakers out of their parents. The
hope that agents would work to renovate rural social institutions, however, was soon dashed. Although there were agents who engaged in
various endeavors of social reformation, particularly school consolidation and curricular reform, most lacked the time, training, or inclination
to involve themselves in such acrimonious and time-consuming affairs.
Early agents were more interested in agricultural club work for boys and
girls, and most made at least some efforts in that direction. But club
work, too, was often time-consuming and only marginally rewarding.
The original idea behind the work was that agricultural clubs would help
advance agricultural education in the schools and would train young
people to be more efficient future farmers and homemakers. Almost
immediately, however, club enthusiasts perceived that club work could
be the "entering wedge" to reform and standardize on-going farming
and animal-breeding practices. Despite the enthusiasm of many agriculturalists and some impressive successes, club work did not usually
measure up to the expectations of its sponsors. Opposition of parents,
apathy of children, poor planning, and the lack of responsible local
leadership combined to limit the effectiveness of club programs. Perhaps
most important, however, agents seldom had the time adequately to
supervise a county-wide club program in addition to their other duties,
and by the mid-teens many counties were hiring special agents to concentrate solely on club work.

Despite their concern with these other areas of rural renovation,
county agents were always most interested in educating adults to be
better farmers and homemakers. This was what they had been trained

SOURCE: Reprinted by permission from *The Resisted Revolution: Urban America and the
Industrialization of Agriculture, 1900–1930* by David B. Danbom © 1979 by the Iowa State
University Press, 2121 South State Avenue, Ames, Iowa 50010.

for, and it was in adult education that agents expended most of their efforts. But before they could begin renovating the economic practices of agriculture, the agents faced the immediate problem of interesting farmers in the work. Businessmen and bankers had been far more important than farmers in securing the passage of the Smith–Lever bill, and few farmers were interested in it at all. Agricultural extension was billed as a reform, but it was the kind of reform in which rural people were least likely to be interested. Reforms traditionally attractive to the countryside were those aimed at the banks, the middlemen, the railroads, and the trusts of the industrial cities, where farmers believed the sources of most of their and the nation's problems could be found. The Smith–Lever measure was an urban thrust which implied that the farmer's problems and the nation's problems arose in the country rather than the city and were directly or indirectly caused by the inefficiency of agriculture. Not surprisingly, many farmers were unenthusiastic about the bill and failed to perceive it as a reform. Many others were not even aware of the Cooperative Extension program. In 1917, H. L. Andrew's first year as agent for Auglaize County, Ohio, his "Narrative Report" repeated a complaint common among agents that "the farmers did not know what a County Agricultural Agent was, and what his duties were. . . . There was no sentiment for the work," and others agreed that initially "the great problem of the county agent . . . is to interest the farmer."

Although rural ignorance was a thorny problem, agents soon discovered that knowledge and interest did not necessarily insure success. Farmers understandably perceived agents as foot soldiers of scientific agriculture, and most were distrustful of science or felt threatened by it when it was applied to farming. There were, of course, a few farmers who were interested in scientific developments in agriculture, and most farm papers and organizations at least publicly contended that farmers could learn from the experts. Yet most agriculturalists agreed with Pennsylvania farmer W. F. McSparren, an articulate critic of scientific agriculturalists and their motives, who commented in the 1908 Vermont *Agricultural Report* that "the farmer who is conservative in his adoption of new notions and practices in the conduct of his business will have fewer disappointments and losses than the enthusiast who accepts all that is spoken and written as true and applicable to all conditions and circumstances." But even when government-sponsored innovations worked in the neighborhood farmers were slow to take them up, contending that the government backed experiments with capital that was not available to the farmer who tried the experiment and failed. And of course experiments did fail, or proved too expensive for farmers to copy, or too complex for farmers to understand, and agents did give bad

advice, all confirming the wisdom of rural conservatism. Extension leaders envisioned an agrarian utopia in which the farmers and the agents
grew just the right crops from tested and improved seed with rational
methods and regularly brought forth two blades of grass where one
grew before. Farmers, however, were skeptical about this dream, and
they were reticent about abandoning proven methods. The old ways
did return crops, and if they did not always yield large crops neither
did they often lead to failure.

So the early agents ran against the coarse grain of rural conservatism. The father of Claude Wickard, Henry A. Wallace's successor as
Secretary of Agriculture, was probably typical. To him, "'book farming'
was taboo. College teachers who taught agriculture were considered
crackpots. Farm magazines were read with extreme skepticism. . . . And
all of this was capped by an inherent rural antagonism to change of any
sort." As local representatives of scientific agriculture, agents and those
who cooperated with them were often the victims of much derision.
"The average boy 10 or 12 years old on the farm knows what . . . will
grow and what will not grow. If farmers had to depend on what these
college boys preach they would be worse off than they are now," hooted
a Nebraska farmer who reflected a popular rural prejudice against sci
entific agriculture and its missionaries.

Already distrustful of scientific agriculture, farmers were bound to
be offended when people they believed lacked practical experience presumed to suggest different methods of farming. "We now have a young
man saddled on us, at an expense of about $3000 per year, to tell us
how to farm,—while he sits in a livery rig to do it," complained a
Minnesota farmer, and an Ohio man noted that farm advice was particularly "irritating because said in a patronizing way." Farm women
were even more offended than their husbands by outside advisers, and
they gave home demonstration agents a particularly frosty reception. "I
am wondering what the Department of Agriculture proposes to do for
us," was one Wisconsin farm wife's bitter question. "Perhaps they may
send out some city women to teach us how to cook. We will resent that."
Her warning was quickly confirmed by home agents who received sharp
rebuffs in their attempts to enlighten farm women on the modern ways
of diet, health, sanitation, cooking, cleaning, canning, and child rearing.
Of course, demonstration officials and agents believed that rural people
had an incorrect conception of the function of extension, and they strove
hard, as one agent put it, "to eradicate the idea that I was here to show
each individual farmer how to farm." Agents wanted farmers to believe
that their job was to help rather than instruct them, but the fact that
their job was in large part to instruct farmers impeded the agents' propaganda campaign.

Farmers also distrusted agents because they believed that the aim of extension was to get farmers to increase production. Leery of the USDA's motives, one California farm wife stated the case succinctly:

> There is a question in my mind as to what to 'aid' us farm women means to you. Is it that you wish us to increase farm production, and are contemplating sending us a lot of pamphlets to 'help' us make more butter, raise more vegetables, supply the markets with more eggs than we do now . . .? If that is what your 'service' to us would imply, I decline it with thanks.

She was, of course, largely right. "The aim of the whole agricultural movement is to maintain and increase the production of the land," admitted prominent agricultural scientist E. H. Jenkins in 1914, and the principal measure of county agent success was whether and how much the agent had increased local production. This was, of course, because the motive behind extension was to raise farmers to a level of efficiency at which they would adequately supplement the industrial nation by supplying it with large amounts of cheap, standardized, and improved produce. "What the Government has done to help the farmer has been done still more to help the city man . . . ," noted Gifford Pinchot in 1918. "What the city man wanted was cheap food. Therefore, what was done for the farmer was directed almost without exception toward helping or inducing him to grow cheap food." The problem encountered by the agents was that most rural experience ran against the wisdom of increasing production. Farmers were no more concerned with the well-being of city people than city people were concerned with the farmers' well-being, and they were understandably distrustful of urban calls for increased production. "Farmers are constantly admonished that the most important economic demand of the time is greater production," noted the *Farm Journal* in January, 1914, "but there is nothing in past experience to indicate to the farmer that he would benefit by an increase in his crops." Throughout the nation farmers noted that big crops usually lost money, and they wondered why the federal government was so intent on increasing production. The Grange and others warned that increasing production was dangerous, particularly when middlemen stood to gain the most from such a move, and at least one farm leader noted that if the government had the right to force down farm prices by making farmers more efficient, it should also make industry more efficient and lower the prices farmers had to pay. Agents answered farmers' objections about extension motives in several ways. Some merely attacked farmers as lazy and claimed it was their duty to produce as much as possible, while others attempted to prove that big crops yielded big returns. The agents' most popular defense was that big crops could pay more when accompanied by innovations lowering the per unit cost of production, and some appealed to rural greed by noting that individual large crops paid well when most farmers continued to grow small ones.

This last appeal touched a basic fact about farmers which agents attempted to exploit. Despite the apparent agreement among farmers that overproduction was dangerous, and despite the warning farm organizations and a few farm papers issued against it, individual farmers could be attracted to large crops, particularly when they believed that their fellows would be returning small ones. Indeed, it was an indication of the farmer's perverse individualism and perhaps even selfishness that the very warnings against overproduction stimulated him to produce more. Assuming that his fellows would not increase production, the individual farmer was often willing to do so to increase his profits. And when the individual wanted to increase production, the agent was there to help him. What this meant was that the same rural individualism which exasperated Country Lifers in so many of their endeavors facilitated the realization of their ends, at least in this case. In the area of production, rural individualism and materialism sometimes eroded the very rational rural distrust of urban motives.

Agricultural production did nudge slowly ahead in the early- and mid-teens. Sometimes tentatively and conservatively, farmers seemed to be responding to the high domestic demand and to the unnatural market situation created by the First World War in Europe. It is doubtful that the county agents had much to do with this increase. They were sometimes available to advise farmers who wished to increase production, but the volition for that increase probably derived more from high prices than from anything government and its emissaries had to say. Indeed, the increase in production came primarily through means the extension personnel found unattractive. Farmers, it seemed, were returning larger crops by bringing more acres into production, not by applying scientific principles to their work. Hence, the per acre yield of most crops remained static, an indication of continuing inefficiency in agricultural production.

Increased production, whether achieved through intensive or extensive means, might still solve the immediate urban problem of expensive food. But the food supply increased much more slowly than the food demand as warring Europeans vied with hungry urbanites for American foodstuffs. By the mid-teens the famine Malthusians had predicted for America seemed to have nearly arrived. Between 1909–1911 and 1914–1916 American wheat imports leaped by over 550 percent and corn imports jumped by nearly 6000 percent. In the same period the market prices of wheat and corn advanced 30.6 percent and 42.1 percent, respectively. Despite their best efforts, extension personnel had been unable significantly to change farm practice or supply cheap food to urban consumers.

Rural skepticism regarding the agents' economic motives was reinforced by the close relationship which emerged between county agents
and local businessmen and bankers. In a sense this nexus cannot be
blamed completely on the agents, for when they arrived in their counties
many were shunned by everyone but merchants and moneylenders. On
the local level businessmen were often the only people interested in
agricultural extension, and the county's share of the agent's salary was
often subscribed by local capitalists or appropriated by the county under
pressure from them. Merchants' enthusiasm for extension rose primarily
because they hoped to make money out of scientific agriculture and
because they shared the agents' assumption that the farmer's problems
were caused by his own inefficiency. On the other hand, agents entering
new counties were always careful to confirm that they had business
support, for businessmen could pressure farmers to cooperate with
agents, and agents who alienated business faced short careers. In their
zeal to garner the aid of commercial people, however, agents sometimes
forgot their original purpose and spent time begging businessmen to
support them or even undertaking business projects, such as helping
erect chambers of commerce and similar commercial clubs. Businessmen
and agents often found themselves in a mutually satisfactory situation,
with the former supporting the latter and urging the farmers to do
likewise, while the agents carried the commercial perspective to the
farmer and were careful that their projects did not offend businessmen.
This relationship was comfortable for agents, bankers, and merchants,
but farmers often regarded it with suspicion. After all, businessmen
expected to profit from the arrangement by selling equipment, lending
money, carrying produce, and by rationalizing markets and distribution.
Not surprisingly, farmers questioned any movement allied with their
age-old economic adversaries in the towns. To attempt to counter rural
suspicions county agents pressed for understanding and cooperation
between businessmen and farmers and joined business in trying to convince farmers that the capitalists had their best interests at heart. As B.
F. Harris, publisher of the *Banker-Farmer*, implied, rural businessmen
and farmers could both be seen as small, interdependent capitalists
struggling to survive in an economy increasingly dominated by massive,
urban-based capital aggregations. But farmers continued to shy from
what Harris saw as a natural alliance of rural enterprisers, and they
continued to distrust merchants and bankers and the county agents
identified with those groups.

A further barrier to extension success was the difficulty county
agents encountered in organizing rural people. Organization was one
of the ends of the industrialization of agriculture, but it was also an
indispensable means to the other end—efficiency. To help achieve the
broad goals of social and economic efficiency, county agents, with the

enthusiastic backing of the USDA, set about organizing the countryside. From the agent's perspective organization was important primarily because it made his job easier. "Christ was able to obtain only 12 Apostles and it might take a County Agent many years to convert an [sic] County of Individuals," noted one Wisconsin agent, "but with groups he need work only a few years and he can revolutionize the agriculture of a county." Others agreed that social efficiency, in the form of organization, was the best means to potential economic efficiency because it facilitated the large-scale teaching of the principles of scientific agriculture. But organization was also important in insuring the survival of the agent. Agents soon realized that organized minorities were often more powerful than unorganized majorities when it came to securing county funds for the work or raising them privately. Organization thus became so important to agents that by the mid-teens in some states agents were not even placed in the field until an organization had been formed in the county to support them.

But getting rural people into organizations was not an easy task. Not only did racial, ethnic, sectarian, political, social, and neighborhood differences and poor communications work against organization, but rural self-reliance and independence also retarded the formation of activist social or economic groups. "The farmer's success has in the past depended very little upon his mental adaptability to other men," concluded Thomas Nixon Carver, the head of the USDA's Rural Organization Service. "He has had to control the forces of nature rather than the forces of society." The farmer's self-centeredness and his consequent social awkwardness and lack of interest in socially or economically active organizations were facts that underlay many of the agents' early difficulties.

Another problem the agents encountered in their organizational endeavors was finding local people willing or able to lead the county farm organizations, but this was just a facet of the larger problem that farmers, as self-conscious individualists, were unwilling to be followers. As the influential *Breeder's Gazette* lamented on September 28, 1916, "The country can never develop and retain great leaders until at least a considerable proportion of the people are willing to be followers." Agents attempted to solve rural leadership problems by doing the work themselves, appointing the leadership and giving it orders, or by turning leadership positions over to the merchants and bankers in the farm bureaus. None of these expedients really worked very well, and they often had the effect of deepening rural suspicions regarding the government, its agents, their motives, and the organizations they formed.

The difficulty agents faced in organizing farmers raises the question of why they failed to utilize the organizations already in existence in the countryside. The fact is that despite the Country Lifers' continual complaints about the lack of organization in the countryside, "When the county agent arrived on the job, he found as a rule fairly good social organizations of farmers in most of the states, such as granges, farmers' clubs, and the like." For a time some agents apparently attempted to work with these groups, but they soon found it necessary to form their own organizations. The reason why agents formed their own groups goes to the very heart of their purpose and underscores the unique and urban-oriented nature of that purpose. The agents formed their own groups because, in the countryside, "There was little, if any, organization, the primary purpose of which was increased efficiency in production." Most existing farm groups did not agree with the agents' assumption that rural problems derived from rural inefficiency. That was an urban—not a rural—analysis of agricultural difficulties, and the agents found it necessary to form entirely new groups which would accept their new point of view. Despite their complaints about rural individualism, then, it was rural distrust of agents' motives and lack of sympathy with their purposes which formed the key impediment to their organizational endeavors.

The organization of farmers, then, was one of the agents' most difficult assignments. But pressing ahead, with the aid of businessmen and a few scientifically oriented farmers, agents bucked widespread opposition and generalized suspicion to bring forth groups—usually called farm bureaus—whose purposes most farmers probably opposed. The organization of a farm bureau did not assure the agent of success, for it faced the same problems of leadership, representation, purpose, and social involvement that plagued every other farm organization, but the farm bureau did give the county agent a place, no matter how tenuous, in at least a segment of the rural community.

This emphasis on rural resistance to county agent work should not obscure the fact that some agents were successful in much of what they tried. Even in the poorest counties a few farmers could be found who believed that agriculture was primitive and should become efficient and organized, farmers who were scientifically oriented and eager for advice. Hence, some agents had extraordinary success and became important men in their counties. Even those who distrusted and disparaged the agents sometimes found them useful. Serious doubts about the motives of the federal government did not prevent farmers from calling on agents for help when they were threatened by individual problems of production. So farmers sometimes found themselves in the paradoxical position of suspecting the extension program in general while using the individual agent. But many agents faced the situation one veteran extension worker remembers from Minnesota, where "the county agent was a sort of 'illegitimate child,' fathered by unhallowed business and left on the farmer's doorstep, certainly not wanted by many farmers."

As outsiders in a hostile or indifferent environment, agents often found it necessary to act with much more circumspection than they might have liked. The lack of rural response to their projects forced many agents to devise plans of work more acceptable to farmers, and sometimes even to undertake projects suggested by farmers themselves, thus often adding detours to what their supervisors hoped would be a straight road to greater productivity. Because of the limitations of their training, most of the agents still undertook projects designed to increase agricultural efficiency. But agents in tenuous situations tended to emphasize projects which would increase economic efficiency (increased income relative to expenditures) rather than those which would increase productive efficiency (increased production per acre and/or per man). This emphasis tended to find more favor with farmers, and a faction within the USDA had long advocated it, but it did not provide the direct route to greater production the Country Lifers had hoped to see in the agents' activities.

Ironically, the farm bureau was one of the things responsible for bringing the county agent under farmer control. Although agents formed bureaus to give themselves an independent power base, they sometimes discovered that they became captives of the bureaus, particularly when the agents, the bureaus, or their members alienated other farmers, other groups of farmers, or, on rare occasions, businessmen. Some agents found themselves tied to the bureaus, taking bureau orders, carrying out bureau programs, and shunned by nonmembers.

Extension personnel generally adjusted to this situation and accepted it. Insofar as the relationship was seen as one in which the "program of work is determined by the local people with the advice of experts," it was accepted and even celebrated as exemplary of the much pursued marriage of democracy and technocracy. Extension Director H. C. Hochbaum of Idaho complained that local people "do not accept the problems we should like them to accept," but even he admitted that "the people have a right to determine for themselves just what help they want." Given the fact that the program was voluntary, this represented a realistic acceptance of the situation. And the agents who were captives of bureaus also seemed to accept the relationship as the correct one.

It would be a mistake, however, to assume that all or even most agents became captives of the farm bureaus, or to assume that capture led to revolutionary alterations in agents' programs. Although M. C. Burritt of Cornell found as late as 1919 that "In a few cases the local associations have too large a control for the best interests of the public partner," he also discovered that "In too many instances the public institutions dominate the partnership." And even where agents were captives, their talents remained oriented towards increasing efficiency and their client groups shared with them the assumption that the problems of agriculture derived from its inefficiency. Although capture sometimes led to shifts in emphasis, the most significant factor in the agent-bureau relationship remained this shared assumption about agriculture and its problems.

But all of these mitigating factors were not clear to the Country Lifers. From their immediate, short-term perspective, the operation of the county agent system in the first years after 1914 was something less than an unqualified success.

It had given its urban supporters hope by advancing to some degree the twin goals of efficiency and organization. But it had not been able socially to regenerate the countryside or adequately to increase the production of food, and in some areas agents had been captured by their own farm bureaus or by existing farm organizations. As developments unfolded, the USDA found itself in a particularly uncomfortable position. Happy that some farmers were beginning to accept the agents, federal officials also shared the concerns of their larger constituency that the agents' presence had not advanced production fast enough to noticeably slow the rise in food prices. By the years of the First World War, then, the agents had achieved a qualified degree of success, but their tenuous hold in the countryside and their declining popularity among urban critics made their situation insecure at best. . . .

THE PRICE OF PROGRESS

By 1930 the industrialization of agriculture was assured. The twenties saw a continuation of the social and economic trends accelerated by World War I, and a consolidation of the economic gains made during the war. The road to an organized and efficient agricultural sector was marked by turns and detours which reflected the difficulties involved in social and economic modification. A major detour came during the New Deal, when the need to keep farmers out of the depression-ridden cities fostered federal initiatives designed to slow, but not reverse, the course of economic change in agriculture. But the basic commitment among policymakers to an organized and efficient agriculture never changed, and when peacetime urban prosperity returned after World War II the agricultural revolution again intensified. Urban America has been able to enjoy the benefits of agricultural industrialization in the post-World War II years because the foundations of an organized and efficient agriculture were laid between 1900 and 1930.

The industrialization of agriculture resulted from both internal and external developments. Certainly, agriculture became more efficient in part because of internal factors. Surplus products had always been needed to pay for land, equipment, and taxes, so the farmer had always been at least part businessman. In addition, the twentieth century saw a tremendous expansion in the production of consumer goods and, for the first time, a systematic exploitation of the rural market. The result was an explosion of rural wants which rapidly became rural needs that had to be filled by surplus production. Because of their organizational difficulties and because of the unique market situation presented by the war, farmers came to believe that increased production was the only practical way to increase profits. Increased production, in turn, demanded increased investment in the factors of production which had to be paid for by further production increases. "Rising standards of rural living, increased use of power and mechanical equipment, and vastly more abundant contact with urban affairs," noted William Jardine in 1929, "all tend to require that farmers place more emphasis than ever before upon cash income." Thus, the relatively self-sufficient and independent yeoman of 1900 had by 1930 become a businessman striving for efficiency, dependent upon suppliers and consumers to fill his needs and wants and buy the surplus he had to sell to survive.

Developments in the countryside were important factors in the industrialization of agriculture, but they were not the only factors. Partially responsible for industrialization was outside pressure from a larger society that demanded an organized and efficient agricultural sector to supplement the emergent industrial nation. The story of American agriculture in the first thirty years of this century is the story of an urban attack on rural isolation, individualism, and self-sufficiency which aimed to make agriculture a socially and economically organized and efficient part of an increasingly interdependent nation. Though they were fascinated by the developments of the urban-industrial nation and attracted by the material fruits it produced, rural people resisted the pressures for industrialization because it was socially and politically irrelevant and economically threatening to them. The ambivalence of rural people regarding social and economic change reflected their ambivalence about rural society. The farmer was proud of his independence and his self-sufficiency, but his life was one of relative social isolation and economic deprivation. He recognized the tangible advantages change might bring, but he also perceived that it potentially endangered his way of life. "The misgiving aroused in the minds of many farmers by the decline of self-sufficiency and the spread of commercialization and urban ways . . . amounted to a perception of the social and economic maladjustment that the modern world was bringing to the countryside," noted Paul H. Johnstone of the USDA in 1940. "The farmer himself, pushed one way by the impact of the new and pulled the other by the persistence of the old, sensed the cultural conflict that was frequently ignored by professional experts, who were for the most part one-sided enthusiasts."

Due to rural resistance, the attempts of the Country Life Movement to industrialize agriculture met with little success until World War I. Then, because of an unnatural market situation and rural patriotism, many farmers became more receptive to increasing production by expanding acreage and by becoming more efficient. When the farmers were ready to produce more, the farsighted missionaries of the New Agriculture were there to help them. Also, the relative scarcity of the means of production and the expansion of federal power gave the USDA tools with which to command the respect and attention of farmers and with which to forward both the short- and long-term goals of industrialization. Not only did the war see an increase in production, it also enhanced the opportunity of the government to identify and organize those farmers most efficient and most likely to see the advantages of industrialization. Even after the war, when the American Farm Bureau Federation became a semiprivate pressure group, the government retained close ties with it because it was the farm organization most likely to agree with the urban critics that agricultural problems derived from rural inefficiency and demanded industrial solutions. The Farm Bureau Federation gave those farmers most likely to embrace the modern ideals of organization and efficiency a say in the future of agricultural policy, but it also gave urban pressure groups and the federal government a means of dealing with agriculture. The war, then, was the pivotal occurrence in the industrialization of agriculture. It unnaturally speeded the attainment of agricultural efficiency to the point where depression resulted, which in turn decreed that the gains of the war would not be lost. Not only did the depression assure the cities of the cheap food that had been their primary goal all along, it also stimulated the sort of sharp competition in agriculture which assured that it would become still more productive. And finally, the war accelerated those trends which were breaking down rural social institutions, leading to increased realization of Country Lifers' social goals.

The way in which public policy was directed toward the industrialization of agriculture presents an interesting case study of decision making in an industrial society. Seldom in the first thirty years of this century was agricultural change made an issue amenable to decision by the representative branches of government. Industrialization was a bureaucratic exercise undertaken by state and federal agencies with the support of powerful private urban interests. Representative bodies never had the opportunity to draw the broad outlines of agricultural policy. They were allowed, as in the Smith-Lever Act, to have some say regarding the vehicles through which the policy would be carried out, but the policy was made by the bureaucracy. And even in the cases where legislation was needed, that legislation was implemented not by the legislators, but by the bureaucrats who had shaped the policy. The extraordinary power of the professional bureaucracy in modern society should not surprise us. Because it can plan, because it has continuity of personnel, and because it is little affected by the vicissitudes of public opinion, the professional bureaucracy assumes an increasingly important role in an industrial society, where planning and continuity are both possible and necessary. In Robert Wiebe's terms, the bureaucracy can provide order in a society which desires it. Most historians who have studied agriculture in the early part of this century have been concerned primarily with agrarian unrest and the legislative response to that unrest. And yet the most significant decisions in agricultural policy came not from legislators and executives concerned with transitory problems, but from bureaucrats, primarily in the USDA, who never lost sight of their goals of rural social and economic efficiency and organization. Elected officials, and the farmers they represented, sometimes pestered the bureaucracy or put it on the defensive, but they were never able significantly to alter its policies.

The crucial role of the bureaucracy in the agricultural revolution conveys the impression that change resulted from the machinations of a manipulative and undemocratic elite. In a discussion of the development of the organizational society in the twentieth century, Samuel P. Hays touches on this manipulation by noting that "while the agents of science and technology have professed to themselves and to the world at large that they were neutral instruments rather than goal-makers, they, in fact were deeply preoccupied with shaping and ordering the lives of other men. They claimed to speak for, to embody, the values of society as a whole." This characterization fits many in the Country Life Movement, including most of the crucial people in the USDA. But it is also true that the Country Lifers did speak for important segments of the larger society, and that their values were values on the ascendancy in that society. The Country Lifers in and out of the bureaucracy embraced the modern, amoral, industrial values of organization and efficiency, values which were coming to replace the Christian and republican values which had traditionally defined the good man and the good society for Americans in the preindustrial age. In enunciating the new values and attempting to put them into practice, the bureaucracy believed it was expressing the goals of the larger society, and it did enjoy the support of important segments of society. Their belief that they represented society and advanced its goals allowed the bureaucrats to resolve the tension between technocracy and democracy, but their virtue was never as clear as they pretended. Certainly, the bureaucracy did not represent the interests of most farmers, and many of its activities were manipulative of and detrimental to those farmers. In short, the activities of the bureaucracy were aimed at aiding one segment of society and expressing its values without regard to the consequences for another segment. It was, in Hays's terms, representing some men by "shaping and ordering the lives of other men." It is an important normative question whether such behavior by government, even if it does express the will of a majority or has the best interests of that majority at heart, is justifiable in a free society.

So in the industrialization of agriculture the bureaucracy played a key role, acting as an equal and perhaps even as a superior branch of government. In their actions the professional bureaucrats were neither impartial brokers nor inert tools. Indeed, they acted as technocrats would be expected to act. They saw more clearly than most of their contemporaries the long-range trends in the United States, and they pursued policies which would make agriculture conform to those trends. Once on their course they did not waver, and even after World War I when most Country Lifers lost interest in agriculture the bureaucrats retained their fidelity to the means and ends of industrialization without questioning the morality of either.

Of course, Country Lifers in and out of the bureaucracy always contended that their innovations would aid farmers, but despite their pronouncements the industrialization of agriculture had never been promoted primarily for the benefit of agriculture. It had been aimed at helping the cities and the nation by making agriculture an efficient supplement to industry which would supply it with cheap food. For the Country Life Movement as a whole the farmer's primary importance was always as a means to urban ends. Of course, there were farmers who accepted the ascendant values of urban society, farmers who benefited from increased efficiency and organization. But, as Paul H. Johnstone noted in 1940, "the deliberate attempts to improve agriculture and rural life . . . have for the most part concentrated their benefits upon the more prosperous element of the farm population. For only the more prosperous ones have been able to take full advantage of modern technology and commercialism." Indeed, the twenties made it abundantly clear that the main beneficiaries of rural progress would be the cities and the efficient farmers, while the majority of rural people would suffer. The developments of the twenties did, however, force some of the critics of agriculture to finally begin seriously to question the effect of rural change on the farmer and the nation.

One of the first facts to become apparent was that as a business proposition farming was made much more difficult as increasing efficiency was demanded. Often beginning his career as an average man with few special talents, the farmer was increasingly "forced to make adjustments to the demands of the new industrial order, involving progressive mechanization, scientific technique, adequate capital, available credit, and dependable markets." Moreover, even if the farmer did improve his techniques he was not assured of greater profits, for his competitors were improving too and he was producing the sorts of inelastic goods which did not sell proportionally better when supplies went up and prices went down. Perhaps most important, the increasing business demands of successful farming effectively closed it as an occupation readily open to the average man. When the century opened farming was the only major occupation in which the citizen without special skills or large capital could achieve the status of an independent businessman. By 1930 farming had been industrialized to the point where the man entering it without expertise or capital faced a marginal existence at best.

Even more important than the effect of industrialization on farming as a business was its effect on the farm as a home. The American farm had always been a home first and a business second. And yet rural change was moving inevitably in the direction of making the farm a business primarily and a home only secondarily, for economic success was unrelated to the emphasis placed on the home or the breadth of its function. Indeed, economic success even seemed to hasten the decline of familial cohesion and to limit the function of the home as a social and educational institution.

Of equal concern was the effect of industrialization on the individual. Rural individualism had long impeded the drive for efficiency and organization, but some commentators began to ask whether it was not preferable to the sort of interdependence in which the farmer was merely a subsidiary adjunct to an industrial society. For years most Country Lifers had assumed a conjunction between efficiency and morality, yet the question was increasingly asked whether these two things were necessarily related. "There is a flood of literature urging the industrialization of agriculture," noted New York pomologist Ulysses P. Hedrick. "From it one would glean that the object of life is to attain efficiency. Some of the happiest, most worthy, and most influential farmers in the State are dreadfully inefficient. A self-respecting freeman is a more desirable citizen than a slave to industry." And Garet Garrett added that, in all its "emphasis on production," the USDA had completely forgotten that the primary value of agriculture came in "spiritual satisfaction, as a manner of living."

Concern over the effect of industrialization on the farmer was inextricably bound to anxiety regarding its effect on the nation as a whole. Declining rural population and rapid urban growth was particularly disturbing to those who feared for the future of the Republic. Urban agrarians had earlier embraced social and economic efficiency as the best means of holding the farmer on the land, but when the twenties proved that efficiency actually sped the demise of the family farm, it was reexamined. "People tell us that the workman at his bench succumbed to the mechanized shop . . . so the small farmer must succumb to the great farming corporation which will operate on a large scale," lamented one commentator. "When we shall have one millionaire farmer driving a thousand people to produce one crop instead of a thousand families producing a thousand crops, then we shall have suffered an irretrievable loss in our national human assets." The farmer would lose, but the nation would be the main loser. Not surprisingly, some critics contended that the nation would suffer because the cities would no longer be assured of a steady stream of moral people; but others steeped in the agrarian myth began to question the very future of liberty and democracy in a nation of few farmers. The survival of republicanism in a nation where more and more people were dependent upon others and upon the government was uncertain. Liberty and democracy had been erected on an agrarian base composed of independent, self-sufficient people. Some believed that as the ancient base disintegrated, danger to the structure it supported increased.

Questions arose, then, particularly during the Great Depression when the basic desirability of industrialism came under some doubt. But it was generally true that, as always, few Americans were anxious to probe what was defined as "progress." In the first years of this century urban America had delineated an agricultural problem. The problem was that an economically and socially inefficient and disorganized agriculture threatened the development, and perhaps even the survival, of a nation increasingly dominated by industry. Unlike other agricultural problems, this one was not perceived as a problem by farmers but only by those among whom urban-industrial concerns were paramount. Thus, the urban-based Country Life Movement was concerned with the social and economic difficulties an unindustrialized agriculture created for urban-industrial society, identifying them as rural problems despite the fact that farmers did not see them as such. What ensued was the classic urban-rural confrontation, in which the countryside resisted urban-born changes which were largely irrelevant to rural people and which attacked traditional practices and institutions. The process of revolutionizing agriculture was an arduous one, and one which continues even today. And yet a combination of factors rising both from within and without rural society had by 1930 given the nation a highly productive agriculture which could only become more efficient. Throughout the period, remarkably few people outside the farm community questioned the means or ends of industrialization. Agricultural renovation was necessary for national economic and social progress and was therefore good. Thus, though the critics' means changed from persuasion to manipulation and compulsion, they were seldom questioned. Nor was their end of an organized and efficient agriculture examined. Seldom did they ask, until it was too late, what the effect of industrialization might be on the farm family. To the social engineers the farmer was first a producer whose virtue was exactly proportional to his productivity. Only secondarily, if at all, was the farmer a human being with hopes and dreams and ideas of his own. Thus, like so many others in the increasingly dehumanized industrial society, the farmer was a means but not an end. Never did his detractors consider shaping the industrial future to him, but only shaping him to the industrial future. Nor were they willing to probe the effect of change on a rural society that, though inefficient in their eyes, had imparted definition and consistency to people's lives for generations. To the critics, rural institutions were merely anachronisms to be modified in the interest of progress, never traditional structures which might be appropriate and valuable. Likewise, the Country Lifers were unwilling to explore the effect of agricultural industrialization on the nation. They were interested in securing an agricultural population at once socially and economically organized and efficient. But the critics failed to wonder if this end would benefit the nation in any but the material sense. Cheap food might be in the national interest, but was it in the national interest effectively to eliminate the primary path to economic independence and proprietorship open to the average American? And once the self-sufficient agrarian base out of which the nation's political principles had grown and upon which its political institutions were erected was altered or diminished, might not the principles and institutions resting thereon face alteration as well? But these doubts were seldom expressed, and were never expressed before the agricultural revolution was well under way. The nation as a whole and the people who wished to renovate agriculture remained unwilling to consider the possible price of progress.

CHAPTER 3

Labor and Monopoly Capital: The Degradation of Work in the Twentieth Century

Harry Braverman

"We are *driven!*" boasts a recent Japanese auto company's commercials, playing on a pun, but also playing on the audience's recognition that our age is one in which everyone feels driven, pushed to a fast tempo in the workplace, pressured to strive for ever higher levels of efficiency. The pressure under which we work has traditionally been viewed as the hallmark of an industrial age, and the breathtaking pace of production is seen as being technologically determined, set less by humans than by the machines we employ supposedly as our servants.

Harry Braverman, however, does not accept this kind of explanation. A Marxist, critical of the capitalist system, Braverman regards the drive for efficiency to be the distinctive hallmark of capitalism and not necessarily a technological imperative of industrialization. He directs our attention to the early years of this century, when we find a fervor for efficiency that almost resembles a religion: The "scientific management" movement that swept factories in the first several decades of the twentieth century. It is the ideas of this movement and its originator, Frederick Winslow Taylor (1856–1915), that Braverman stringently criticizes as the culprits.

In the selection presented here, we find two kinds of material: Taylor's ideas, in his own words, and Braverman's critique of the ideas, which Taylor viewed as scientific observations but which Braverman regards as merely the techniques of capitalists who sought to control production more firmly.

One of the underlying tenets of Taylor's system, and a tenet that Braverman takes issue with, is that the biggest obstacle to obtaining maximum output from a worker is the worker's inclination to purposely adopt a slow pace in the work, which was called "soldiering" in the jargon of the day. In Taylor's story of how he quadrupled the daily workload of Schmidt (a pseudonym), a loader in the Bethlehem Steel yards, we are left to assume that Schmidt's co-workers were working far less than they should have been.

Taylor's innovation was his assignment of what he called "task work" at Bethlehem. "Day work" (payment per day, regardless of output) and "piece work" (payment per physical unit of work completed, regardless of time used) were "old-fashioned," in Taylor's view. His task work combined elements of both: payment was issued per day, but every motion of the worker was supervised and paced by management, so that the tasks defined for the day would be completed.

Braverman criticizes Taylor's success at Bethlehem for several reasons. First, it led to effective cuts in the workers' pay. Taylor paid a sixty percent bonus to Schmidt for his successful sustaining of a much-quickened work pace, but Braverman argues that employers did not continue to pay such a premium. What actually happened, he says, is that the pace was sped up but pay remained the same. Second, Braverman criticizes the fact that only a small number of workers could physically stand up to Taylor's pacing.

The experiment at Bethlehem involved a simple task, loading pig iron. Most disturbing to Braverman was the extension of Taylor's system to more complex work, which had traditionally been carried out by craftsmen and craftswomen. Taylor studied machinists' work, for example, with the aim of learning the optimum variables needed for a given task and making it possible for management, not the machinists, to control the work process.

Taylor's system can be summarized in three principles. First, traditional knowledge is gathered into the hands of management. Braverman refers to this as the "dissociation of the labor process from the skills of the workers." Second, conception of a task is separated from its execution. At first glance, this might appear to mean that the mental aspects of a task are to be separated from the physical, and in a factory setting this would indeed apply. But the principle also could apply to management of work in an office, where all work is apparently "mental," since the conception of a given task can still be separated from its execution. The third and final principle is that management uses its monopoly over knowledge to control each step of the labor process. No detail, however small, is left to the individual worker to decide.

Braverman is concerned about the loss of craftsmanship in the workplace that has been a consequence of acceptance of Taylorian ideas. His concern springs from his own experience working for twelve years as a coppersmith, pipe fitter, and sheet metal worker. Elsewhere in his writings he claims that he harbors no "sentimental attachment" to "archaic" forms of labor, but it is difficult to picture how we could enjoy our abundant creature comforts of today and yet still maintain a craft-based system of production, in which planning and execution remain unified, and individual workers determine the pace of work.

An important assumption made by Braverman, which you may wish to challenge, is that capitalism dictates the particular form of workplace organization. It is capitalism, in Braverman's view, that demands that the division of labor be so refined that any individual worker's given task necessarily must be repetitive, dull, and meaningless; this is what he refers to as "the degradation of work," mentioned in the subtitle of his book. But there are at least a few counterexamples of alternative organizational technologies in use in capitalist systems, such as at the Volvo Auto Assembly Plant in Kalmar, Sweden, where no conventional assembly line exists; workers are organized into teams, which assemble entire systems for each car and which rotate individual assignments for variety.

As for the organization of work in socialist societies, at least to date the record shows few examples of successful factories organized along lines that differ significantly from their capitalist counterparts. But it is important to search for comparisons across countries, as well as back through time, for only when we look at the broadest range of work experiences can we judge the relative importance of technology in shaping the texture of modern work.

SCIENTIFIC MANAGEMENT

Control has been the essential feature of management throughout its history, but with Taylor it assumed unprecedented dimensions. The stages of management control over labor before Taylor had included, progressively: the gathering together of the workers in a workshop and the dictation of the length of the working day; the supervision of workers to ensure diligent, intense, or uninterrupted application; the enforcement of rules against distractions (talking, smoking, leaving the workplace, etc.) that were thought to interfere with application; the setting of production minimums; etc. A worker is under management control when subjected to these rules, or to any of their extensions and variations. But Taylor raised the concept of control to an entirely new plane when he asserted as an *absolute necessity for adequate management the dictation to the worker of the precise manner in which work is to be performed.* That management had the right to "control" labor was generally assumed before Taylor, but in practice this right usually meant only the general setting of tasks, with little direct interference in the worker's mode of performing them. Taylor's contribution was to overturn this practice and replace it by its opposite. Management, he insisted, could be only a limited and frustrated undertaking so long as it left to the worker any decision about the work. His "system" was simply a means for management to achieve control of the actual mode of performance of every labor activity, from the simplest to the most complicated. To this end, he pioneered a far greater revolution in the division of labor than any that had gone before.

Taylor created a simple line of reasoning and advanced it with a logic and clarity, a naive openness, and an evangelical zeal which soon won him a strong following among capitalists and managers. His work began in the 1880s but it was not until the 1890s that he began to lecture, read papers, and publish results. His own engineering training was limited, but his grasp of shop practice was superior, since he had served a four-year combination apprenticeship in two trades, those of pattern-maker and machinist. The spread of the Taylor approach was not limited to the United States and Britain; within a short time it became popular in all industrial countries. In France it was called, in the absence of a suitable word for management, "l'organisation scientifique du travail" (later changed, when the reaction against Taylorism set in, to "l'organisation rationnelle du travail"). In Germany it was known simply as *rationalization;* the German corporations were probably ahead of everyone else in the practice of this technique, even before World War I.

Taylor was the scion of a well-to-do Philadelphia family. After preparing for Harvard at Exeter he suddenly dropped out, apparently in rebellion against his father, who was directing Taylor toward his own profession, the law. He then took the step, extraordinary for anyone of his class, of starting a craft apprenticeship in a firm whose owners were social acquaintances of his parents. When he had completed his apprenticeship, he took a job at common labor in the Midvale Steel Works, also owned by friends of his family and technologically one of the most advanced companies in the steel industry. Within a few months he had passed through jobs as clerk and journeyman machinist, and was appointed gang boss in charge of the lathe department.

In his psychic makeup, Taylor was an exaggerated example of the obsessive-compulsive personality: from his youth he had counted his steps, measured the time for his various activities, and analyzed his motions in a search for "efficiency." Even when he had risen to importance and fame, he was still something of a figure of fun, and his appearance on the shop floor produced smiles. . . .

The forms of management that existed prior to Taylorism, which Taylor called "ordinary management," he deemed altogether inadequate. . . . His descriptions of ordinary management bear the marks of the propagandist and proselytizer: exaggeration, simplification, and schematization. But his point is clear:

> Now, in the best of the ordinary types of management, the managers recognize frankly that the . . . workmen, included in the twenty or thirty trades, who are under them, possess this mass of traditional knowledge, a large part of which is not in the possession of management. The management, of course, includes foremen and superintendents, who themselves have been first-class workers at their trades. And yet these foremen and superintendents know, better than any one else, that their own knowledge and personal skill falls far short of the combined knowledge and dexterity of all the workmen under them. The most experienced managers frankly place before their workmen the problem of doing the work in the best and most economical way. They recognize the task before them as that of inducing each workman to use his best endeavors, his hardest work, all his traditional knowledge, his skill, his ingenuity, and his good-will—in a word, his "initiative," so as to yield the largest possible return to his employer.

As we have already seen from Taylor's belief in the universal prevalence and in fact inevitability of "soldiering," he did not recommend reliance upon the "initiative" of workers. Such a course, he felt, leads to the surrender of control: "As was usual then, and in fact as is still usual in most of the shops in this country, the shop was really run by the workmen and not by the bosses. The workmen together had carefully planned just how fast each job should be done." In his Midvale battle, Taylor pointed out, he had located the source of the trouble in the "ignorance of the management as to what really constitutes a proper day's work for a workman." He had "fully realized that, although he was foreman of the shop, the combined knowledge and skill of the workmen who were under him was certainly ten times as great as his own." This, then, was the source of the trouble and the starting point of scientific management.

We may illustrate the Taylorian solution to this dilemma in the same manner that Taylor often did: by using his story of his work for the Bethlehem Steel Company in supervising the moving of pig iron by hand. This story has the advantage of being the most detailed and circumstantial he set down, and also of dealing with a type of work so simple that anyone can visualize it without special technical preparation. We extract it here from Taylor's *The Principles of Scientific Management:*

> One of the first pieces of work undertaken by us, when the writer started to introduce scientific management into the Bethlehem Steel Company, was to handle pig iron on task work. The opening of the Spanish War found some 80,000 tons of pig iron placed in small piles in an open field adjoining the works. Prices for pig iron had been so low that it could not be sold at a profit, and therefore had been stored. With the opening of the Spanish

War the price of pig iron rose, and this large accumulation of iron was sold. This gave us a good opportunity to show the workmen, as well as the owners and managers of the works, on a fairly large scale the advantages of task work over the old-fashioned day work and piece work, in doing a very elementary class of work.

The Bethlehem Steel Company had five blast furnaces, the product of which had been handled by a pig-iron gang for many years. This gang, at this time, consisted of about 75 men. They were good, average pig-iron handlers, were under an excellent foreman who himself had been a pig-iron handler, and the work was done, on the whole, about as fast and as cheaply as it was anywhere else at that time.

A railroad switch was run out into the field, right along the edge of the piles of pig iron. An inclined plank was placed against the side of a car, and each man picked up from his pile a pig of iron weighing about 92 pounds, walked up the inclined plank and dropped it on the end of the car.

We found that this gang were loading on the average about 12½ long tons per man per day. We were surprised to find, after studying the matter, that a first-class pig-iron handler ought to handle between 47 and 48 long tons per day, instead of 12½ tons. This task seemed to us so very large that we were obliged to go over our work several times before we were absolutely sure that we were right. Once we were sure, however, that 47 tons was a proper day's work for a first-class pig-iron handler, the task which faced us as managers under the modern scientific plan was clearly before us. It was our duty to see that the 80,000 tons of pig iron was loaded on to the cars at the rate of 47 tons per man per day, in place of 12½ tons, at which rate the work was then being done. And it was further our duty to see that this work was done without bringing on a strike among the men, without any quarrel with the men, and to see that the men were happier and better contented when loading at the new rate of 47 tons than they were when loading at the old rate of 12½ tons.

Our first step was the scientific selection of the workman. In dealing with workmen under this type of management, it is an inflexible rule to talk to and deal with only one man at a time, since each workman has his own special abilities and limitations, and since we are not dealing with men in masses, but are trying to develop each individual man to his highest state of efficiency and prosperity. Our first step was to find the proper workman to begin with. We therefore carefully watched and studied these 75 men for three or four days, at the end of which time we had picked out four men who appeared to be physically able to handle pig iron at the rate of 47 tons per day. A careful study was then made of each of these men. We looked up their history as far back as practicable and thorough inquiries were made as to the character, habits, and the ambition of each of them. Finally we selected one from among the four as the most likely man to start with. He was a little Pennsylvania Dutchman who had been observed to trot back home for a mile or so after his work in the evening, about as fresh as he was when he came trotting down to work in the morning. We found that upon wages of $1.15 a day he had succeeded in buying a small plot of ground, and that he was engaged in putting up the walls of a little house for himself in the morning before starting to work and at night after leaving. He also had the reputation of being exceedingly "close," that is, of placing a very high value on a dollar. As one man whom we talked to about him said, "A penny looks about the size of a cart-wheel to him." This man we will call Schmidt.

The task before us, then, narrowed itself down to getting Schmidt to handle 47 tons of pig iron per day and making him glad to do it. This was done as follows. Schmidt was called out from among the gang of pig-iron handlers and talked to somewhat in this way:

"Schmidt, are you a high-priced man?"

"Vell, I don't know vat you mean."

"Oh yes, you do. What I want to know is whether you are a high-priced man or not."

"Vell, I don't know vat you mean."

"Oh, come now, you answer my questions. What I want to find out is whether you are a high-priced man or one of these cheap fellows here. What I want to find out is whether you want to earn $1.85 a day or whether you are satisfied with $1.15, just the same as all those cheap fellows are getting."

"Did I vant $1.85 a day? Vas dot a high-priced man? Vell, yes, I vas a high-priced man."

"Oh, you're aggravating me. Of course you want $1.85 a day—every one wants it! You know perfectly well that that has very little to do with your being a high-priced man. For goodness' sake answer my questions, and don't waste any more of my time. Now come over here. You see that pile of pig iron?"

"Yes."

"You see that car?"

"Yes."

"Well, if you are a high-priced man, you will load that pig iron on that car to-morrow for $1.85. Now do wake up and answer my question. Tell me whether you are a high-priced man or not."

"Vell—did I got $1.85 for loading dot pig iron on dot car to-morrow?"

"Yes, of course you do, and you get $1.85 for loading a pile like that every day right through the year. That is what a high-priced man does, and you know it just as well as I do."

"Vell, dot's all right. I could load dot pig iron on the car to-morrow for $1.85, and I get it every day, don't I?"

"Certainly you do—certainly you do."

"Vell, den, I vas a high-priced man."

"Now, hold on, hold on. You know just as well as I do that a high-priced man has to do exactly as he's told from morning till night. You have seen this man here before, haven't you?"

"No, I never saw him."

"Well, if you are a high-priced man, you will do exactly as this man tells you to-morrow, from morning till night. When he tells you to pick up a pig and walk, you pick it up and you walk, and when he tells you to sit down and rest, you sit down. You do that right straight through the day. And what's more, no back talk. Now a high-priced man does just what he's told to do, and no back talk. Do you understand that? When this man tells you to walk, you walk; when he tells you to sit down, you sit down, and you don't talk back at him. Now you come on to work here to-morrow morning and I'll know before night whether you are really a high-priced man or not."

This seems to be rather rough talk. And indeed it would be if applied to an educated mechanic, or even an intelligent laborer. With a man of the mentally sluggish type of Schmidt it is appropriate and not unkind, since it is effective in fixing his attention on the high wages which he wants and away from what, if it were called to his attention, he probably would consider impossibly hard work. . . .

Schmidt started to work, and all day long, and at regular intervals, was told by the man who stood over him with a watch, "Now pick up a pig and walk. Now sit down and rest. Now walk—now rest," etc. He worked when he was told to work, and rested when he was told to rest, and at half-past five in the afternoon had his 47½ tons loaded on the car. And he practically never failed to work at this pace and do the task that was set him during the three years that the writer was at Bethlehem. And throughout this time he averaged a little more than $1.85 per day, whereas before he had never received over $1.15 per day, which was the ruling rate of wages at that time in Bethlehem. That is, he received 60 per cent higher wages than were paid to other men who were not working on task work. One man after another was picked out and trained to handle pig iron at the rate of 47½ tons per day until all of the pig iron was handled at this rate, and the men were receiving 60 per cent more wages than other workmen around them.

The merit of this tale is its clarity in illustrating the pivot upon which all modern management turns: the control over work through the control over the *decisions that are made in the course of work.* Since, in the case of pig-iron handling, the only decisions to be made were those having to do with a time sequence, Taylor simply dictated that timing and the results at the end of the day added up to his planned day-task. As to the use of money as motivation, while this element has a usefulness in the first stages of a new mode of work, employers do not, when they have once found a way to compel a more rapid pace of work, continue to pay a 60 percent differential for common labor, or for any other job. Taylor was to discover (and to complain) that management treated his "scientific incentives" like any other piece rate, cutting them mercilessly so long as the labor market permitted, so that workers pushed to the Taylorian intensity found themselves getting little, or nothing, more than the going rate for the area, while other employers—under pressure of this competitive threat—forced their own workers to the higher intensities of labor.

Taylor liked to pretend that his work standards were not beyond human capabilities exercised without undue strain, but as he himself made clear, this pretense could be maintained only on the understanding that unusual physical specimens were selected for each of his jobs:

> As to the scientific selection of the men, it is a fact that in this gang of 75 pig-iron handlers only about one man in eight was physically capable of handling 47½ tons per day. With the very best of intentions, the other seven out of eight men were physically unable to work at this pace. Now the one man in eight who was able to do this work was in no sense superior to the other men who were working on the gang. He merely happened to be a man of the type of the ox,—no rare specimen of humanity, difficult to find and therefore very highly prized. On the contrary, he was a man so stupid that he was unfitted to do most kinds of laboring work, even. The selection of the man, then, does not involve finding some extraordinary individual, but merely picking out from among very ordinary men the few who are especially suited to this type of work. Although in this particular gang only one man in eight was suited to doing the work, we had not the slightest difficulty in getting all the men who were needed—some of them from inside the works and others from the neighboring country—who were exactly suited to the job.

Taylor spent his lifetime in expounding the principles of control enunciated here, and in applying them directly to many other tasks: shoveling loose materials, lumbering, inspecting ball bearings, etc., but particularly to the machinist's trade. He believed that the forms of control he advocated could be applied not only to simple labor, but to labor in its most complex forms, without exception, and in fact it was in machine shops, bricklaying, and other such sites for the practice of well-developed crafts that he and his immediate successors achieved their most striking results.

From earliest times to the Industrial Revolution the craft or skilled trade was the basic unit, the elementary cell of the labor process. In each craft, the worker was presumed to be the master of a body of traditional knowledge, and methods and procedures were left to his or her discretion. In each such worker reposed the accumulated knowledge of materials and processes by which production was accomplished in the craft. The potter, tanner, smith, weaver, carpenter, baker, miller, glassmaker, cobbler, etc., each representing a branch of the social division of labor, was a repository of human technique for the labor processes of that branch. The worker combined, in mind and body, the concepts and physical dexterities of the specialty: technique, understood in this way, is, as has often been observed, the predecessor and progenitor of science. The most important and widespread of all crafts was, and throughout the world remains to this day, that of farmer. The farming family combines its craft with the rude practice of a number of others, including those of the smith, mason, carpenter, butcher, miller, and baker, etc.

The apprenticeships required in traditional crafts ranged from three to seven years, and for the farmer of course extends beyond this to include most of childhood, adolescence, and young adulthood. In view of the knowledge to be assimilated, the dexterities to be gained, and the fact that the craftsman, like the professional, was required to master a specialty and become the best judge of the manner of its application to specific production problems, the years of apprenticeship were generally needed and were employed in a learning process that extended well into the journeyman decades. Of these trades, that of the machinist was in Taylor's day among the most recent, and certainly the most important to modern industry.

As I have already pointed out, Taylor was not primarily concerned with the advance of technology (which, as we shall see, offers other means for direct control over the labor process). He did make significant contributions to the technical knowledge of machine-shop practice (high-speed tool steel, in particular), but these were chiefly by-products of his effort to study this practice with an eye to systematizing and classifying it. His concern was with the control of labor at any given level of technology, and he tackled his own trade with a boldness and energy which astonished his contemporaries and set the pattern for industrial engineers, work designers, and office managers from that day on. And in tackling machine shop work, he had set himself a prodigious task.

The machinist of Taylor's day started with the shop drawing, and turned, milled, bored, drilled, planed, shaped, ground, filed, and otherwise machine- and hand-processed the proper stock to the desired shape as specified in the drawing. The range of decisions to be made in the course of the process is—unlike the case of a simple job, such as the handling of pig iron—by its very nature enormous. Even for the lathe alone, disregarding all collateral tasks such as the choice of stock, handling, centering and chucking the work, layout and measuring, order of cuts, and considering only the operation of turning itself, the range of possibilities is huge. Taylor himself worked with twelve variables, including the hardness of the metal, the material of the cutting tool, the thickness of the shaving, the shape of the cutting tool, the use of a coolant during cutting, the depth of the cut, the frequency of regrinding cutting tools as they became dulled, the lip and clearance angles of the tool, the smoothness of cutting or absence of chatter, the diameter of the stock being turned, the pressure of the chip or shaving on the cutting surface of the tool, and the speeds, feeds, and pulling power of the machine. Each of these variables is subject to broad choice, ranging from a few possibilities in the selection and use of a coolant, to a very great number of effective choices in all matters having to do with the thickness, shape, depth, duration, speed, etc. Twelve variables, each subject to a large number of choices, will yield in their possible combinations and permutations astronomical figures, as Taylor soon realized. But upon these decisions of the machinist depended not just the accuracy and finish of the product, but also the pace of production. Nothing daunted, Taylor set out to gather into management's hands all the basic information bearing on these processes. He began a series of experiments at the Midvale Steel Company, in the fall of 1880, which lasted twenty-six years, recording the results of between 30,000 and 50,000 tests, and cutting up more than 800,000 pounds of iron and steel on ten different machine tools reserved for his experimental use. His greatest difficulty, he reported, was not testing the many variations, but holding eleven variables constant while altering the conditions of the twelfth. The data were systematized, correlated, and reduced to practical form in the shape of what he called a "slide rule" which would determine the optimum combination of choices for each step in the machining process. His machinists thenceforth were required to work in accordance with instructions derived from these experimental data, rather than from their own knowledge, experience, or tradition. This was the Taylor approach in its first systematic application to a complex labor process. Since the principles upon which it is based are fundamental to all advanced work design or industrial engineering today, it is important to examine them in detail. And since Taylor has been virtually alone in giving clear expression to principles which are seldom now publicly acknowledged, it is best to examine them with the aid of Taylor's own forthright formulations.

First Principle

"The managers assume . . . the burden of gathering together all of the traditional knowledge which in the past has been possessed by the workmen and then of classifying, tabulating, and reducing this knowledge to rules, laws, and formulae. . . ." We have seen the illustrations of this in the cases of the lathe machinist and the pig-iron handler. The great disparity between these activities, and the different orders of knowledge that may be collected about them, illustrate that for Taylor— as for managers today—no task is either so simple or so complex that it may not be studied with the object of collecting in the hands of management at least as much information as is known by the worker who performs it regularly, and very likely more. This brings to an end the situation in which "Employers derive their knowledge of how much of a given class of work can be done in a day from either their own experience, which has frequently grown hazy with age, from casual and unsystematic observation of their men, or at best from records which are kept, showing the quickest time in which each job has been done."

It enables management to discover and enforce those speedier methods and shortcuts which workers themselves, in the practice of their trades or tasks, learn or improvise, and use at their own discretion only. Such an experimental approach also brings into being new methods such as can be devised only through the means of systematic study.

This first principle we may call the *dissociation of the labor process from the skills of the workers*. The labor process is to be rendered independent of craft, tradition, and the workers' knowledge. Henceforth it is to depend not at all upon the abilities of workers, but entirely upon the practices of management.

Second Principle

"All possible brain work should be removed from the shop and centered in the planning or laying-out department. . . ." Since this is the key to scientific management, as Taylor well understood, he was especially emphatic on this point and it is important to examine the principle thoroughly.

In the human, as we have seen, the essential feature that makes for a labor capacity superior to that of the animal is the combination of execution with a conception of the thing to be done. But as human labor becomes a social rather than an individual phenomenon, it is possible— unlike in the instance of animals where the motive force, instinct, is inseparable from action—to divorce conception from execution. This dehumanization of the labor process, in which workers are reduced almost to the level of labor in its animal form, while purposeless and unthinkable in the case of the self-organized and self-motivated social labor of a community of producers, becomes crucial for the management of purchased labor. For if the workers' execution is guided by their own conception, it is not possible, as we have seen, to enforce upon them either the methodological efficiency or the working pace desired by capital. The capitalist therefore learns from the start to take advantage of this aspect of human labor power, and to break the unity of the labor process.

This should be called the principle of the *separation of conception from execution*, rather than by its more common name of the separation of mental and manual labor (even though it is similar to the latter, and in practice often identical). This is because mental labor, labor done primarily in the brain, is also subjected to the same principle of separation of conception from execution: mental labor is first separated from manual labor and, as we shall see, is then itself subdivided rigorously according to the same rule.

The first implication of this principle is that Taylor's "science of work" is never to be developed by the worker, always by management.

This notion, apparently so "natural" and undebatable today, was in fact vigorously discussed in Taylor's day, a fact which shows how far we have traveled along the road of transforming all ideas about the labor process in less than a century, and how completely Taylor's hotly contested assumptions have entered into the conventional outlook within a short space of time. Taylor confronted this question—why must work be studied by the management and not by the worker himself; why not *scientific workmanship* rather than *scientific management?*—repeatedly, and employed all his ingenuity in devising answers to it, though not always with his customary frankness. In *The Principles of Scientific Management,* he pointed out that the "older system" of management

> makes it necessary for each workman to bear almost the entire responsibility for the general plan as well as for each detail of his work, and in many cases for his implements as well. In addition to this he must do all of the actual physical labor. The development of a science, on the other hand, involves the establishment of many rules, laws, and formulae which replace the judgment of the individual workman and which can be effectively used only after having been systematically recorded, indexed, etc. The practical use of scientific data also calls for a room in which to keep the books, records, etc., and a desk for the planner to work at. Thus all of the planning which under the old system was done by the workman, as a result of his personal experience, must of necessity under the new system be done by the management in accordance with the laws of the science; because even if the workman was well suited to the development and use of scientific data, it would be physically impossible for him to work at his machine and at a desk at the same time. It is also clear that in most cases one type of man is needed to plan ahead and an entirely different type to execute the work.

The objections having to do with physical arrangements in the workplace are clearly of little importance, and represent the deliberate exaggeration of obstacles which, while they may exist as inconveniences, are hardly insuperable. To refer to the "different type" of worker needed for each job is worse than disingenuous, since these "different types" hardly existed until the division of labor created them. As Taylor well understood, the possession of craft knowledge made the worker the best starting point for the development of the science of work; systematization often means, at least at the outset, the gathering of knowledge which *workers already possess.* But Taylor, secure in his obsession with the immense reasonableness of his proposed arrangement, did not stop at this point. In his testimony before the Special Committee of the House of Representatives, pressed and on the defensive, he brought forth still other arguments:

> I want to make it clear, Mr. Chairman, that work of this kind undertaken by the management leads to the development of a science, while it is next

to impossible for the workman to develop a science. There are many work-men who are intellectually just as capable of developing a science, who have plenty of brains, and are just as capable of developing a science as those on the managing side. But the science of doing work of any kind cannot be developed by the workman. Why? Because he has neither the time nor the money to do it. The development of the science of doing any kind of work always required the work of two men, one man who actually does the work which is to be studied and another man who observes closely the first man while he works and studies the time problems and the motion problems connected with this work. No workman has either the time or the money to burn in making experiments of this sort. If he is working for himself no one will pay him while he studies the motions of some one else. The management must and ought to pay for all such work. So that for the workman, the development of a science becomes impossible, not because the workman is not intellectually capable of developing it, but he has neither the time nor the money to do it and he realizes that this is a question for the management to handle.

Taylor here argues that the systematic study of work and the fruits of this study belong to management for the very same reason that ma-chines, factory buildings, etc., belong to them; that is, because it costs labor time to conduct such a study, and only the possessors of capital can afford labor time. The possessors of labor time cannot themselves afford to do anything with it but sell it for their means of subsistence. It is true that this is the rule in capitalist relations of production, and Taylor's use of the argument in this case shows with great clarity where the sway of capital leads: Not only is capital the property of the capitalist, but *labor itself has become part of capital.* Not only do the workers lose control over their instruments of production, but they must now lose control over their own labor and the manner of its performance. This control now falls to those who can "afford" to study it in order to know it better than the workers themselves know their own life activity.

But Taylor has not yet completed his argument: "Furthermore," he told the Committee, "if any workman were to find a new and quicker way of doing work, or if he were to develop a new method, you can see at once it becomes to his interest to keep that development to himself, not to teach the other workmen the quicker method. It is to his interest to do what workmen have done in all times, to keep their trade secrets for themselves and their friends. That is the old idea of trade secrets. The workman kept his knowledge to himself instead of developing a science and teaching it to others and making it public property." Behind this hearkening back to old ideas of "guild secrets" is Taylor's persistent and fundamental notion that the improvement of work methods by workers brings few benefits to management. Elsewhere in his testimony, in discussing the work of his associate, Frank Gilbreth, who spent many years studying bricklaying methods, he candidly admits that not only *could* the "science of bricklaying" be developed by workers, but that it undoubtedly *had been:* "Now, I have not the slightest doubt that during the last 4,000 years all the methods that Mr. Gilbreth developed have many, many times suggested themselves to the minds of bricklayers." But because knowledge possessed by workers is not useful to capital, Taylor begins his list of the desiderata of scientific management: "First. The development—by the management, not the workmen—of the science of bricklaying." Workers, he explains, are not going to put into execution any system or any method which harms them and their workmates: "Would they be likely," he says, referring to the pig-iron job, "to get rid of seven men out of eight from their own gang and retain only the eighth man? No!"

Finally, Taylor understood the Babbage principle better than anyone of his time, and it was always uppermost in his calculations. The purpose of work study was never, in his mind, to enhance the ability of the worker, to concentrate in the worker a greater share of scientific knowledge, to ensure that as technique rose, the worker would rise with it. Rather, the purpose was to cheapen the worker by decreasing his training and enlarging his output. In his early book, *Shop Management,* he said frankly that the "full possibilities" of his system "will not have been realized until almost all of the machines in the shop are run by men who are of smaller calibre and attainments, and who are therefore cheaper than those required under the old system."

Therefore, both in order to ensure management control and to cheapen the worker, conception and execution must be rendered separate spheres of work, and for this purpose the study of work processes must be reserved to management and kept from the workers, to whom its results are communicated only in the form of simplified job tasks governed by simplified instructions which it is thenceforth their duty to follow unthinkingly and without comprehension of the underlying technical reasoning or data.

Third Principle

The essential idea of "the ordinary types of management," Taylor said, "is that each workman has become more skilled in his own trade than it is possible for any one in the management to be, and that, therefore, the details of how the work shall best be done must be left to him." But, by contrast: "Perhaps the most prominent single element in modern scientific management is the task idea. The work of every workman is fully planned out by the management at least one day in advance, and each man receives in most cases complete written instructions, describing in detail the task which he is to accomplish, as well as the means to be used in doing the work. . . . This task specifies not only what is to be done, but how it is to be done and the exact time allowed for doing it. . . . Scientific management consists very largely in preparing for and carrying out these tasks."

In this principle it is not the written instruction card that is important. Taylor had no need for such a card with Schmidt, nor did he use one in many other instances. Rather, the essential element is the systematic pre-planning and pre-calculation of all elements of the labor process, which now no longer exists as a process in the imagination of the worker but only as a process in the imagination of a special management staff. Thus, if the first principle is the gathering and development of knowledge of labor processes, and the second is the concentration of this knowledge as the exclusive province of management—together with its essential converse, the absence of such knowledge among the workers—then the third is the *use of this monopoly over knowledge to control each step of the labor process and its mode of execution*. . . .

CHAPTER 4

America by Design: Science, Technology, and the Rise of Corporate Capitalism

David F. Noble

The achievements of our early inventors continue to be celebrated; our schoolchildren become well acquainted with Ben Franklin, Alexander Graham Bell, and Thomas Alva Edison. Schools do not do nearly as well, however, in acquainting us with the individuals who were responsible for more recent inventions. Who invented the jet engine? The semiconductor chip? The dry photocopy machine? The laser? We can speculate about reasons for the anonymity of contemporary invention. The sheer proliferation of important inventions, of course, makes it unlikely that we will learn and retain the details of the origins of each. But more important, the nature of invention itself has changed—today it is often the result of a group research process, conducted under corporate (or military) sponsorship. The lone individual who, on her or his own, is struck by the flash of genius and comes up with an idea for a better mousetrap does not represent the typical inventor of today.

David Noble's book helps us understand how invention in America was transformed in the early twentieth century. His book concerns the process by which science and engineering in America fell under the direction of large corporations. According to Noble, scientists and engineers lost their independence to corporate capitalism, and so, too, did America's universities and engineering schools. Noble is sharply critical of this process, and he holds that these events have had a critical influence in determining what things have been invented and what things have not been, what technology has been utilized and what has not.

In this selection, the author discusses the growing corporate dominance over patents during the late nineteenth and early twentieth centuries. Originally, the framers of the Constitution called specifically for a system that conferred exclusive rights to actual inventors for their discoveries. The life of a patent eventually was set at seventeen years, and both inventor and society enjoyed mutual benefits. The inventor had a period of time in which to commercialize and profit from the invention, and society benefitted from rapid commercialization of a new discovery, yet could also look forward to a limited term before the inventor's monopoly expired and the invention became part of the public domain.

The largest corporations seized upon patents, one legal form of monopoly, as loopholes for escaping the antitrust laws, especially in the period between 1900 and 1929. Through licensing agreements, mergers with smaller patent-holding companies, patent purchases, and internal research, the largest companies could attain complete dominance of a particular technology. Noble illustrates the pattern with the examples of AT&T (telephony and radio) and of General Electric (incandescent lamps).

In cases where no single company could wrest a dominant position, a truce would be called and a patent-pool set up, in which the biggest players agreed to cooperate with each other. Such pools were legal, and they greatly strengthened the dominant positions of the participants, making it difficult for weaker rivals to compete or for new competitors to enter the market.

Another tactic corporations adopted in the battle over patents was massive spending on internal research. The scale of the largest of these efforts was astonishing. Between 1916 and 1935, AT&T's spending on research and engineering far exceeded the total operating budget for Harvard University during the same period.

If individual inventors remained unmoved by the "carrot" of corporate research funding, they were beaten down by the corporation's "stick" of unending legal suits for alleged patent infringement, suits which, even when utterly lacking in merit, were expensive to defend against. For survival, independent inventive souls joined corporations, but were compelled to sign away patent right claims for any work done while an employee. Noble regards this compulsory signing away of rights as a condition of employment to be the ultimate perversion of the patent system envisioned by the Constitution's framers.

In the corporate research laboratories, the nature of invention itself was also changed. Team research supplanted the individual's idiosyncratic quest; specialized assignments broke the invention process into small pieces. Corporate invention, we are told, now took place on an "assembly line," and presumably was drained of the pleasure of serendipitous discovery that had attracted the independent inventor of old.

But we might ask whether the author has overlooked some advantages found in the team approach to research that redound to the benefit of the country as well as to the sponsoring company. A concentration of expensive apparatus is affordable if utilized by the many, but would be prohibitively costly if acquired for the use of a lone individual. And the collaborative approach of a team also offers advantages in intramural stimulation and the convenience of expert consultation with colleagues on the spot.

THE CORPORATION AS INVENTOR

Patent-Law Reform and Patent Monopoly

> *The patent system was established, I believe, to protect the lone inventor. In this it has not succeeded. . . . The patent system protects the institutions which favor invention.*

—E. F. W. Alexanderson

When he intimated his opinion of those particular inventions and discoveries which had most facilitated other inventions and discoveries, Abraham Lincoln included, along with the art of writing and printing and the discovery of America, "the introduction of patent laws." For it was these laws, he explained to his Springfield audience in 1860, that had "added the fuel of interest to the fire of genius," by conferring the protection of monopoly over an invention to the "true inventor" exclusively, thereby directly rewarding the inventive spirit. When Lincoln's oft-quoted words were inscribed in stone above the doors of the new Patent Office in 1932, however, they no longer conveyed either the intent or the *de facto* practice of the system administered therein. "In his day, Abraham Lincoln could well say that 'the patent system added the fuel of interest to the power of genius,'" one observer concluded after the hearings of the Senate Patent Committee in 1949. "Today it would be more correct to say that the patent system adds another instrument of control to the well-stocked arsenal of monopoly interests . . . it is the corporations, not their scientists, that are the beneficiaries of patent privileges." The mass of evidence of the corporate use of patents to

---

SOURCE: From *America by Design: Science, Technology, and the Rise of Corporate Capitalism* by David F. Noble. Copyright © 1977 by David F. Noble. Reprinted by permission of Alfred A. Knopf, Inc.

circumvent antitrust laws which was collected in the testimony before the Temporary National Economic Commission in the early 1930s prompted another writer to concur. "It would require more than twenty years of Rip Van Winkle oblivion to events of this world," he wrote, "to miss the fact that the overwhelming proportion of significant inventions now come out of scientific laboratories, and that these . . . are institutions which have largely if not wholly removed—by deliberate intent—the pecuniary reward for the inventor."

These latter-day critics were but echoing the warnings of those who had much earlier witnessed the transformation of the patent system. Within a half-century after Abraham Lincoln offered his glowing evaluation of it, the American patent system had undergone a dramatic change; rather than promoting invention through protection of the inventor, the patent system had come to protect and reward the monopolizer of inventors, the science-based industrial corporations. "It is well known that patents in the United States are bought up in large numbers for the purpose of suppressing competition," one commentator observed in the *Iron Trade Review* of 1915. He noted that the monopoly of an industry by means of patent control constituted a "monopoly of monopolies" and "a patent on the very industry" as a whole. Such control of patents, he warned, with the resultant capacity for direction and suppression of invention itself, "strangles the sciences and the useful arts, and contributes liberally to illegitimate commercial schemes." As it gave rise to "monopoly of monopolies," the patent system gradually fostered the corporate control of the process of invention itself and thus facilitated the commercially expedient retardation, as well as promotion, of invention.

The framers of the Constitution, who formulated the basis of the American patent system, had deliberately sought to avert this possibility, and in doing so they had departed significantly from the practice of their time. Letters patent had been issued in England as early as the sixteenth century, in accordance with the principles of unwritten common law, and the first patent statute was passed by Parliament in 1623. The patent system of England, however—which was adopted by the colonies and the states under the Articles of Confederation—was geared toward the promotion of new industries by granting monopolies to importers of inventions and processes as well as to inventors themselves. Reward went to the "man who introduces or improves the manufacture and not alone the man who originated the improvement." The writers of the American Constitution, however, "introduced a radically new idea into the view of the function and scope of a patent system."

A proposition placed before the Constitutional Convention would have empowered Congress "to establish public institutions, rewards and immunities for the promotion of agriculture, commerce and manufactures." This was rejected by the convention in favor of one which authorized Congress "to promote the progress of science and the useful arts, by securing for limited times to authors and inventors the exclusive right to their respective writings and discoveries." "For the first time in the world," the nation's most prominent patent lawyer explained in 1909, "the framers of our constitution laid the entire stress . . . on the recognition and reward of inventive thought." The system thus focused upon the inventor, who alone could receive a patent for a particular invention and subsequently either work under it, sell it in part or whole, or grant licenses for its exclusive or nonexclusive use. Legally, there was "no limit to the inventor's absolute control of the thing" covered by the patent, and it could be a process, method, machine, manufacture, a composition of matter, or any improvement of them; ideas or principles, mathematical formulae, laws of nature, or philosophical abstractions "not embodied in concrete form," however, could not be patented. "In the beginning," one student of the patent system has written, "it was easy to fit the definition of invention to the simple economic conditions that prevailed. Any invention was ordinarily the creation of one individual. . . . Our patent system was designed to stimulate the individual to invent by giving him the right to exclude others from making, using, and selling his invention." In the eyes of the designers of the system, moreover, the rewards were not the inventor's alone; the patent right was "really a just reward for service rendered to the community," and the community benefited as well.

The patent system was created for the mutual benefit of the inventor and of society, to which he disclosed his invention in return for patent protection. It was assumed that the patentee would certainly develop the invention for commercial use under the protection granted him; indeed, "the whole historical background of granting monopolies for promoting the progress of the useful arts gives no sanction to the suppression of inventions." As a means of fostering commercial progress, the patent system did not "sanction a monopoly of kindred or competitive patents or a restraint of trade other than in the particular thing which the patent covers"; the Constitution, moreover, provided for patent monopolies "for limited times only" to guard against any long-term restraint of trade. Between 1790 and the latter part of the nineteenth century, however, the role of patents in American commercial development underwent significant changes. These affected the methods by which patents were issued and to whom they were issued, as well as their use once granted.

The first United States Patent Law, of 1790, was administered by Thomas Jefferson and his colleagues under very strict standards, and relatively few patents were issued. Three years later a more relaxed system was adopted whereby "anyone who swore to the originality of his invention and paid the stipulated fees could secure a patent," its validity being decided by the courts. In 1836 this second law was repealed and a Patent Office was created. The 1836 Patent Act "marked the beginning of our present patent system," based upon the "examination system" involving scrutiny of each patent application. The fantastic growth in the number of patent applications thereafter had begun by the end of the nineteenth century to place a great strain upon the rather meager resources and small staff of examiners in the Patent Office.

Praising the admirable foresight of the Founding Fathers, Frederick Fish explained how they had "adopted a new theory that men are encouraged to invent by the certainty of reward; if the fire of genius is fed with the fuel of interest, the industries will take care of themselves." As a former president of AT&T and a GE counsel, Fish could declare with assurance that the industries had done so, and in ways never imagined by the framers of the Constitution: "as business units became larger, patent-owning corporations supplanted inventors in the exploitation of patents." The inventor, the original focus of the patent system, tended increasingly to "abandon" his patent in exchange for corporate security; he either sold or licensed his patent rights to industrial corporations or assigned them to the company of which he became an employee, bartering his genius for a salary. In addition, by means of patent control gained through purchase, consolidation, patent pools, and cross-licensing agreements, as well as by regulated patent production through systematic industrial research, the corporations steadily expanded their "monopoly of monopolies." Although the first patent pool, among manufacturers of sewing-machine parts, was established as early as 1856, it was not until the end of the century that corporations clearly became the dominant factor in patent exploitation. In 1885 twelve percent of patents were issued to corporations; by 1950 "at least three-fourths of patents [were] assigned to corporations." The change in the focus of the patent system, from the protection of the inventor to the protection of the corporation which either employed the inventor or purchased his patents, was succinctly phrased by E.F.W. Alexanderson, a Swedish immigrant who became one of GE's early leading research engineers. "The patent system was established, I believe," he said, "to protect the lone inventor. In this it has not succeeded . . . the patent system protects the institutions which favor invention."

The growth of the corporations, and the intensification of their control through trusts, holding companies, mergers and consolidations, and the community of interest created by intercorporate shareholding and interlocking directorates generated a counterdevelopment within American society: the antitrust laws. The patent system which conferred legal monopolies to inventors came into increasing conflict with antimonopoly legislation as corporations replaced lone inventors as the primary holders of patents, and used patents to create monopolies. The conflict surfaced in court interpretations of the patent monopolies in the light of the Sherman and, after 1914, the Clayton antitrust acts. Section three of the latter explicitly declared the illegality of monopoly based upon sales regulation of patented machinery and products, and was largely a response to the monopoly held by the United Shoe Machinery Company. The interpretations of corporate patent practice varied from court to court, but the judicial history of patent monopolies falls roughly into three periods. Between the signing of the Constitution and the first decade of the twentieth century there was either disregard for, or approval of, monopolies based upon patents. For two decades thereafter there was a gradual tightening of restrictions, although as late as 1926, in an important precedent-setting case involving General Electric, the Supreme Court "emphasized the right of a patent owner to license manufacturers with restrictions as to price." In the third period, beginning "about 1940, indifference and leniency in general gave way to a more aggressive prosecution and court decisions and decrees which reflect[ed], as never before, the purpose of the Sherman Act." The period under examination here, that of 1900–1929, was one of comparatively little judicial restriction of corporate patent monopoly and the market control it made possible. The tremendous strides made along these lines in this period, in addition, were of such proportions as to render subsequent judicial and legislative efforts to check corporate monopoly through patent control too little too late.

The novel American patent system, designed to protect the inventor by granting him a monopoly over his creations, had by the turn of the century fostered the development of "institutions" that demanded a controlled promotion of the "progress of science and the useful arts," one that conformed to the exigencies of corporate stability and prosperity. The science-based industries, based upon patent monopolies from the outset, thus sought to redefine the patent system as yet another means to corporate ends. In particular, they aimed to bend the system in ways which would enable them to circumvent the antitrust laws. Their efforts included intercorporate agreements, industrial research and regulated patent production, and reform of the patent-system apparatus. Edwin J. Prindle, a mechanical engineer and patent lawyer, was active in all three areas.

In numerous articles Prindle outlined the means of securing patent monopolies to bypass the antitrust laws; methods of securing patents from inventors, and employee-inventors; and the legislative means of streamlining the patent system along corporate lines. An early member of the American Patent Law Association, which was founded in 1897, Prindle pursued a career which involved him in countless court cases in the defense of patent-holding corporations, and provided him with the opportunity of formulating, along with Frederick Fish and the other members of the National Research Council Committee on Patents, the bill which authorized the revamping of the Patent Office in the early 1920s. In a widely read series in *Engineering Magazine* in 1906, entitled *Patents as a Factor in a Manufacturing Business,* Prindle clearly spelled out the possible uses of the patent system for purposes of corporate monopoly. In offering his suggestions, he indicated that they arose out of his own experience at the patent bar and as a practicing engineer, as well as from the successful experiences of the pioneers in such undertakings: Bell Telephone, GE, Westinghouse, and the United Shoe Machinery Company.

> Patents are the best and most effective means of controlling competition. They occasionally give absolute command of the market, enabling their owner to name the price without regard to cost of production. . . . Patents are the only legal form of absolute monopoly. In a recent court decision the court said, "within his domain, the patentee is czar . . . cries of restraint of trade and impairment of the freedom of sales are unavailing, because for the promotion of the useful arts the constitution and statutes authorize this very monopoly."
>
> The power which a patentee has to dictate the conditions under which his monopoly may be exercised has been used to form trade agreements throughout practically entire industries, and if the purpose of the combination is primarily to secure benefit from the patent monopoly, the combination is legitimate. Under such combinations there can be effective agreements as to prices to be maintained . . . ; the output for each member of the combination can be specified and enforced . . . and many other benefits which were sought to be secured by trade combinations made by simple agreements can be added. Such trade combinations under patents are the only valid and enforceable trade combinations that can be made in the United States.

Prindle proceeded to outline methods of prolonging monopolies and expanding them through ownership of auxiliary patents. "If a patent can't be secured on a product," he suggested that "it should be secured on processes for making the product." And "if none of these ways is feasible, it should be considered whether or not the product cannot be tied up in some way with a patent on some other product, process, or machine." As a patent lawyer, Prindle understood that "a patent is valid only when granted in the name of the inventor," and he emphasized the importance for corporations, of securing the patent rights of their employees. He alluded to a long series of cases in which corporations were unsuccessful in their attempts to gain control of patented inventions of their employees because they had failed to contract with them specifically for such privileges. He thus strongly argued that

> It is desirable to have a contract with every employee who is at all likely to make inventions which relate to the business of the employer . . . the courts will sustain such contracts, even though they contain no further

provision for return for the inventions than the payment of the ordinary salary. . . .

Prindle was aware that he was deliberately subverting the intent of the patent system. In citing cases where employees had refused to give up "their rights" guaranteed by the Constitution, he emphasized the importance of using "psychology" to obtain the patent rights of employees, acknowledging thereby that "rights" were in fact being lost, and that at least some employees were fully aware of it. "The difficulty of inducing the employees to sign such a contract," he noted, "will be reduced if the officers of the company will set the example by signing such a contract." Quite clearly, Prindle understood that what he was proposing—the compulsory signing of employment contracts which automatically assigned employee patent rights to the employer— amounted to confiscation, and something that neither he nor his readers would have liked to have happen to them. Prindle thus acknowledged that the signing of the contract by the officer to set a "reasonable" example was in fact reasonable for the officer alone, since for him it was "a mere matter of form, as [he] is frequently a man who is either not inventive or one who is glad to take his returns in the form of dividends from the stock." For corporate employees in the science-based industries, however, this matter of form would become standard and compulsory procedure.

The methods outlined by Prindle, were in part the fruit of long experience in the electrical industry. The largest corporations of the industry—GE and AT&T—had years before mapped out the patent territory and begun to refine their tactics for mastering it. Although patent-control measures became widespread in the automobile, rubber, steel, chemical, and other industries, the stories of AT&T and GE in particular and the electrical industry in general provide the earliest examples of how they operated.

AT&T, as has already been noted, was incorporated as the consolidation of the various Bell System interests in 1900. By that time the Bell System had already substantially occupied the field, since it had been "successful in every contest involving the original patents." Having anticipated the expiration of those patents, the Bell companies had, as President Theodore N. Vail phrased it, "surrounded the business with all the auxiliary protection that was possible." As a result, AT&T found itself in an excellent position to stifle and harass competitors through patent-infringement suits. The success of these procedures was explained by an AT&T patent lawyer:

> It appears to me that the policy of bringing suit for infringement on apparatus patents is an excellent one because it keeps the concerns which attempt opposition in a nervous and excited condition since they never

know where the next attack may be made, and since it keeps them all the time changing their machines and causes them ultimately, in order that they may not be sued, to adopt inefficient forms of apparatus.

Among the patents secured by AT&T were those which underlay AT&T's monopoly of long-distance telephony—Michael Pupin's patent on loading coils, and the Cooper-Hewitt patents on the mercury-arc repeater. AT&T purchased Pupin's rights in 1900 and secured exclusive domestic rights for the Cooper-Hewitt patents in 1907. AT&T was also able to obtain the rights for Lee De Forest's three-element vacuum tube in 1913, half a dozen years after it was first patented. By gaining control of the De Forest invention, "which is the heart of radio broadcasting, wireless reception and amplification of long-distance conduction of electrical waves, whether used in radio, telegraph, or telephone communications," AT&T secured a key position in the nascent radio industry.

Between the time of its organization as the Bell Patent Association in 1875 and the creation of the Federal Communications Commission in 1934, the Bell System "remained free from federal regulation." By 1935, through licensing agreements, mergers, purchases, and research, it had increased its patent holdings from two—the original Bell patents—to 9,255, which included some of the most important inventions in telephony and radio. In addition, through the various radio-patent pool agreements of the 1920s, AT&T had effectively consolidated its position relative to the other giants in the industry. An FCC investigation of the telephone industry and extensive study of the patent system led Floyd L. Vaughan to conclude that

> By amassing thousands of patents on inventions in the whole field of communication . . . American Telephone dominates the telephone and also controls "the exploitation of potentially competitive and emerging forms of communication." It thus excludes others from its field and avoids being excluded by them. Would-be rivals may enter and remain only as licensees under restricted conditions. It pre-empts for itself new frontiers of technology for exploitation in the future and, in the meantime, protects what is already developed. It keeps itself in a commanding position for the exchange of patent rights. In short, it employs patents to maintain its dominance . . . in communication.

The experience of General Electric was similar. GE was formed in 1892 as the consolidation of the assets, and especially patents, of the Edison and Thomson-Houston interests. Beginning in 1896 with the establishment of the GE-Westinghouse Board of Patent Control, "the first important example of collusion in acquiring patent rights," GE "followed the conscious policy of funnelling into its control all patents held by its licensees and touching any phase of the [incandescent lighting] industry." A Tariff Commission report on incandescent electric lamps indicated that "since that time, through the purchase and consolidation of numerous companies, through the purchase of patents and through its own research organization . . . GE has acquired most of the important patents covering electric lamps, their parts and machinery and processes for making them." Subsequent court decisions further indicated the success of GE's patent policies. A favorable Supreme Court decision of 1926 noted that GE's control of the patents covering the manufacture of tungsten filaments, and the use of gas in light bulbs to increase light intensity—Just and Hanaman (1912), Coolidge (1913), and Langmuir (1916)—"secure to GE the monopoly of the manufacture of making, using, and vending" of the means of "the making of the modern electric light." In 1949 a considerably less sympathetic New Jersey District court found that GE's "offensive of patents" had led to a situation in which "it individually monopolizes patents employed in the incandescent electric lamp industry." In the opinion of the court,

> General Electric's apparently impregnable position was a formidable barrier to anyone who contemplated entering the lamp manufacturing field and this, coupled with the knowledge that it controlled the manufacture of lamp bases, lamp manufacturing machinery, along with a tight block on the supply of glass, created a situation sufficient to deter entry. The link to unlawful monopoly is apparent from the fact that upon expiration of the lawful patent monopoly in 1933, there was no new entry into the field.

The individual policies of GE and AT&T were carefully designed to gain and prolong monopolies over patents vital to their industry. Toward this end, they employed such methods as incomplete disclosure of information in patent applications, the use of trademarks, the outright suppression or delayed introduction of patented apparatus, the compulsory assignment of employee patents to the company, and the deliberate production of auxiliary patents. Perhaps of greater significance than all of these combined, however, were the agreements through which the efforts of individual companies were coordinated in the interest of all, and insulation from national and international competition in particular fields was secured. The radio-patent pool agreements of the 1920s, among such odd bedfellows as AT&T, GE, RCA, United Fruit, American Marconi, and Westinghouse, provide an illuminating example.

By the beginning of World War I, a number of companies had arrived at a stalemate with regard to radio development, due to mutual patent interferences. During the war, when the government guaranteed to protect the companies from infringement suits, research in radio proceeded at a rapid pace. The close of the war, however, brought with it a renewed deadlock. "Ownership of the various patents pertaining to vacuum tubes and circuits by different concerns prevented the manufacture of an improved tube for radio use." In addition to domestic competition, there was a very real possibility that control over radio might be secured by the British Marconi Company, which was trying at the time to obtain rights to the necessary GE-controlled Alexanderson alternator.

In light of this threat to American supremacy of the airwaves, Woodrow Wilson and a number of armed-forces representatives prevailed upon GE to withhold the necessary patent rights and set up instead an American-owned company to control radio. In late 1919, GE thus established the Radio Corporation of America; it purchased the stock of the British-controlled American Marconi Company and transferred its assets, along with the Alexanderson and other GE-owned patents, to RCA. The industry-wide impasse nevertheless remained, and "the only solution to the conflicts was to declare a truce: get together and draw up an agreement defining the rights of the various squatters on the frontiers of science." The truce was declared between AT&T and GE in the license agreement of July 1, 1920, and within the following year, through collateral agreements, the other companies in the patent conflict joined the radio-patent pool.

The consolidation of radio patents, which numbered some thousand, "divided up the telephone, electrical, and radio fields and established supremacy in each through exclusive licenses." At the same time, the agreements kept all who were not party to them out of the radio field. AT&T, for example, retained "exclusive rights under its patents in two-way telephone services, both wire and radio," while GE and Westinghouse received "all patents held by AT&T and RCA relating to the lamp industry, in return for their patent rights applicable to telephone and radio." "Each one, in effect, kept and obtained the patent rights in its particular field." Perhaps the best description of the intent behind the agreements was offered by J.E. Otterson, general commercial manager of the Western Electric Company, in 1927:

> The regulation of the relationship between two such large interests as the AT&T Company and the GE Company and the prevention of invasion of their respective fields is accomplished by mutual adjustment within . . . the "no-man's land" lying between . . . where the offensive of the parties as related to these competitive activities is recognized as a natural defense against invasion of the major fields. Licenses, rights, opportunities and privileges in connection with these competitive activities are traded off against each other and inter-changed in such manner as to create a proper balance and satisfactory relationship between the parties in the major fields.

"The contract," Otterson explained,

> is an example of the character of arrangement that may develop out of an effort on the part of two large interests to avoid an invasion of their respective fields and a destructive conflict of interests. It was through trading

off rights in connection with these competitive activities that an adjustment between the two interests was reached and the two major fields left intact.

What was actually left "intact," of course, were not the fields themselves, but rather the corporate control over the fields, the "spheres of influence" of the large companies. Through agreements like this within the country, and the establishment of international cartels to regulate the global "field," the large corporations of the electrical industry sought to dominate not only markets for their products but the manufacture of those products as well. Their "Napoleonic concept of industrial warfare, with inventions and patents as the soldiers of fortune," served them well, as the court's evaluation of GE activities in 1949 clearly indicated; GE, it concluded,

> paced its industrial achievements with efforts to insulate itself from competition. It developed a tremendous patent framework and sought to stretch the monopoly acquired by patents far beyond the intendment of those grants. It constructed a great network of agreements and licenses, national and international in scope, which had the effect of locking the door of the United States to any challenge to its supremacy . . . arising from business enterprise indigenous to this country or put forth by foreign manufacturers.

The corporate control of patents by means of intercorporate agreement was inextricably coupled with the research arm of the "patent offensive." As a basis for systematic patent production and monopolization, industrial research played an increasingly important role, from the turn of the century on, in the development of bargaining power for such agreements. Although the primary objective of research was to find solutions to immediate technical problems, another objective was "to anticipate inventive trends and take out patents to keep open the road of technical progress and business expansion."

Scientific research at AT&T began in 1907 when J.J. Carty became chief engineer and brought into the domain of AT&T the research which his predecessor Hammond Hayes had left to the students of MIT and Harvard College. The growing importance of research within the company was highlighted against the backdrop of a severe economic downturn. J.P. Morgan formally took over the reins of the company the same year and, in the wake of the bankers' panic, called for greater efficiency and increased standardization of equipment in order to cut costs. By 1910, however, there were 192 engineers doing development work under Carty, with annual expenditures of half a million dollars. By 1916 this number had increased to 959, with expenditures of $1.5 million, and by 1930 AT&T was spending $25 million for research. Between 1916 and 1935 the engineering department of the Western Electric Company and the Bell Laboratories combined, spent $250 million on engineering and research; this sum, as one historian of AT&T has estimated, "far exceeded the total operating budget, for instance, of Harvard University for the same period."

A major impetus behind this expansion was the early recognition, by Carty and his superiors, of the potentials of wireless, coupled with the need to develop a "repeater" for long-distance telephony. That the promise of radio, in terms of AT&T prosperity and competitive strength, was a prime motivation behind the rapid expansion of research was suggested by Carty himself, in a memorandum of 1909.

> At the present time scientists in Germany, France, Italy, and a number of able experimenters in America are at work upon the problem of wireless telephony. While this branch of the art seems at present to be rather remote in its prospects of success, a most powerful impetus would be given to it if a suitable telephone repeater [vacuum-tube amplifier] were available. Whoever can supply and control the necessary telephone repeater will exert a dominating influence in the art of wireless telephony when it is developed. The lack of such a repeater . . . and the number of people at work upon [it] . . . created a situation which may result in some of these outsiders developing a telephone repeater before we have obtained one ourselves, unless we adopt vigorous measures from now on. A successful telephone repeater . . . might put us in a position of control with respect to the art of wireless telephony should it turn out to be a factor of importance

Work on this problem preoccupied the research branch of the company as soon as it was established. As Frank Jewett, head of the Bell Labs, recalled in 1932, "it was early clear to the AT&T Company . . . that a full, thorough, and complete understanding of radio must be had at all times if the art of telephony . . . was to be advanced and the money invested in that service safeguarded." Such research, in which the prime motivation was commercial dominance through patent offense and defense, proved quite successful; although AT&T had to purchase the De Forest vacuum tube, the research done in its laboratories provided a still greater bargaining position in the radio-patent pool agreements after the war. Perhaps the clearest statement of intent behind such industrial research was provided by J.E. Otterson in his memorandum of 1927:

> A primary purpose of the AT&T Company is the defense and maintenance of its position in the telephone field. . . . Undertakings and policies must be made to conform to the accomplishment of this purpose. The AT&T Company is surrounded by potentially competitive interests which may in some manner or degree intrude upon the telephone field. The problem is to prevent this intrusion.
>
> It seems obvious that the best defense is to continue activities in "no man's land" and to maintain such a strong engineering, patent, and commercial situation in connection with these competitive activities as to always have something to trade against the accomplishment of other parties. . . . It seems essential to . . . maintain an active offensive in the "no man's land"

lying between it and potentially competitive interests. . . . The nearer the trading can be carried to the major field of our competitors the more advantageous the trading position we are in. . . . Ability to stop the owner of a fundamental and controlling patent from realizing the full fruits of his patent by the ownership of necessary secondary patents may easily put one in position to trade where money alone may be of little value.

While research and patent warfare on the part of large "institutions which promote invention" provided them with a competitive edge in their bargaining among themselves, it completely overwhelmed the independent inventor whom the patent system was originally designed to protect. Lone inventors could either try to fight for their rights within "no man's land" or join the dominant forces which occupied the fields around it. Out of frustration and survival instinct, they increasingly flocked to corporate employment in exchange for security and abandoned their patent privileges in the process.

When inventors decided to fight for an independent position, they were faced with frustration on three fronts: trying to obtain a patent; trying to sell or develop it; and trying to see it put to use after it was sold. Because of the conscious policies and extensive resources of the larger corporations, lone inventors were obstructed at every turn. As Floyd Vaughan observed,

> If the inventor sells his patent rights at all, it is usually for a lump sum rather than for royalties. If he develops his own invention, which is seldom, he must seek the capital of others. Most of his inventions are never sold or developed at all. In any case he usually receives little or nothing. As the obstacles of the inventor have grown, patents, to an increasing extent, have stimulated him through delusion rather than reward.

Obtaining the patent was perhaps hardest of all:

> It is a common practice, especially of large companies well-financed and equipped with technicians and patent lawyers, to take out every possible patent in their fields and thus block any would-be intruder. If an outsider seeks a patent in this domain, he must find out in some instances about hundreds of patents on kindred ideas and avoid them. Creative minds may be compelled to spend more time in obtaining or avoiding patents than in solving a problem.

By means of interference and infringement suits, the corporations were well able, and equally inclined, to harass patent applicants and cause them to abandon their claims. According to the president of the Thomas Edison Company, Edison himself had spent more money in obtaining patents, litigating them, and preventing infringements than he had received from them. Lee De Forest, while successful in defending his claims and selling some rights to AT&T for a sizable sum, was pushed into bankruptcy as a result of other patent litigation. In 1917 B.A. Behrend awarded Nikola Tesla the coveted Edison Medal of the AIEE with the words: "Were we to seize and to eliminate from our industrial world the results of Mr. Tesla's work, the wheels of industry would cease to turn, our electrical cars and trains would stop, our towns would be dark, our mills would be dead and idle. Yea, so far-reaching is this work that it has become the warp and woof of industry." Tesla himself, however, as Alexanderson later recalled, "was a frustrated inventor and had to spend his old age in impoverished retirement."

Edison, De Forest, and Tesla, whatever the cost, had been able to translate their inventions into commercial developments, and to sell them. The majority of inventors were less fortunate. While perhaps able to obtain patents, they were unable either to interest investors in them or to sell them to established companies. At a House committee hearing on the pooling of patents, one witness testified that "the greater the contribution, the more certain is it to be denied recognition by the entrenched corporations and their servile laboratory staffs. And the lack of such recognition . . . [in part] explains the shameful spectacle of every single one of the world's great inventions having been forced to be idle until outside competition had forced their adoption despite the cunning and conspiracy of the great corporations in that field—and often only after the inventor was no longer here to receive his due reward." The individual corporations were usually resistant to significant change simply because of their established dominance in the field; moreover, pooling agreements often required the sharing of important new developments with competitors, and this canceled out any motivation of an individual company to invest in a new invention. For similar reasons, if a company did invest in something new, more than likely it was in order to obtain some future bargaining advantage, and thus the immediate use of the patent was suppressed.

In 1912 Louis Brandeis concluded that "these great organizations are constitutionally unprogressive." But in terms of the human cost of independent invention, they were more than that. "Edwin Armstrong," wrote E.F.W. Alexanderson, "was one of the few inventors who made a fortune out of his patents. But the end was sad. His patent litigations got so on his nerves that he committed suicide." Alexanderson, whose alternator was the focus of attention and GE's strongest asset in the radio-patent pool agreements, provided some insight into life outside the pool. Both he and Owen D. Young, chairman of the board of GE, had received a number of threatening letters. "An inventor," Alexanderson recalled, "claimed that he had a patent which was being infringed by my patent. The threats were directed primarily at Mr. Young. He was the head of the company. The letter writer said that it costs only a thousand dollars to get somebody killed and he had the thousand dollars. General Electric had to fight a suit in patent court to prove that his patent . . . was different from my patent. I never met the inventor . . . but those who did meet him said that he was a simple, mild-mannered man."

The frustrations of independent invention led the majority of inventors into the research laboratories of the large corporations; in the process, invention itself was transformed. "Team research in the laboratory of the large corporation has largely displaced the inventive activity of the individual. The assembly line of invention, like that of manufacturing, is dominant today. The improvements of various workmen and technicians are put together, under the guidance of lawyers and business managers, so that patents can be acquired which will provide for dominating a field of production more completely." Inventors became employees in corporations to spare themselves the hardships of going it alone. Their patents were thereby handled by corporation-paid patent lawyers and their inventions were made commercially viable at company expense. Corporate employment thus eliminated the problem of lawsuits, and in addition provided well-equipped laboratories, libraries, and technical assistance for research. The nature of their actual work, however, changed. "Work was often done under high pressure. The employee-inventor was expected to direct his efforts along lines in accord with the company's commercial policies and not to spend time fooling around with any interesting idea that appealed to him. He was expected to produce results of definite commercial value and not to take too long about it." The "collectivization" of invention done in the research laboratory presupposed the specialization of each task: "company inventors were usually organized into departments or sections; they were assigned definite projects to work on and problems to solve," and the various efforts so assigned were assembled only by management.

The new role of the "employee-inventor" further reinforced the changes in the *de facto* patent system that had created him. By incorporating within them the material and human means of technological development, the large corporations effectively eliminated the threat of "outsiders." As one historian of the patent system, Floyd Vaughan, observed, "By controlling the only market for improvement patents and by controlling the factory operations of laboratories where new and pertinent ideas were most likely to occur, a company could command the stream of inventive thought."

According to the United States patent system, "no one except the true inventor can obtain a valid patent." By employing the technical experts capable of producing inventions, the corporations were also obtaining the legally necessary vehicles for the accumulation of corporate patents. At first a number of corporations provided limited compensation to an employee for a patentable invention; GE, for example, rewarded the lucky employee with a dollar. In time, however, employees became required to assign all patent rights to their employer, as part of the employment contract, in return for their salaries. In addition, "to make sure that an employee does not conceive a bright idea and then leave the company to develop it on his own . . . most such employment contracts contain a 'trailing clause.' By means of such a clause, the company can claim inventions not only during employment but also for a period—a year, for instance—after employment has ended." The Bell Laboratories at first compensated employees for patents beyond their salaries, but, as Frank Jewett explained, such incentive allowances encouraged the worker to work for himself rather than his employer, and in competition with his coworkers.

> The incentive was to get as many patents that could pass the Patent Office as possible. An invention was made. It could be covered by one strong patent or it could be covered by a dozen minor patents. It was to the company's advantage to have one strong patent, but it was to the employee's advantage to have a dozen minor patents . . . It created a situation where men would not work with each other . . . yet the problem which was before us was a problem which required team action; . . . so some way had to be found to get over that.

The Bell System's solution was the one Prindle had suggested in 1906: the elimination of patent reward for employees. The "fuel of interest" was completely divorced from the "fire of genius" and, as one writer put it, "the heroic age of American invention" had come to an end. It was true that corporate employees, while no longer able to exploit the fruit of their own inventiveness, nor even to exercise that inventiveness fully, were nevertheless able to eat regularly, "a consideration not to be sneered at." In addition, they had, as Alexanderson noted, "a safe and resourceful place to work." The safety within the corporation, however, had been achieved at the expense of, and in deliberate violation of, whatever safety there was outside it. More and more independent inventors were forced to "abandon their patents" as they would a sinking ship, and seek refuge on the shore from which the ship was being bombarded. After hearing the testimony before the Senate Patent Committee in the early 1940s, Bernhard J. Stern concluded that

> Genius is not nourished, for when the research worker joins the staff of an industrial laboratory, he relinquishes his right to patent the fruits of his researches to the corporation which employs him. If, on the other hand, he remains a member of that almost extinct tribe of solo inventors, he is usually powerless to compete with the industrial giants that control credit, technological facilities, and the markets, and he is generally unable to develop his patent in the face of the expenses of infringement suits.

"No one except the true inventor can obtain a valid patent," Frederick Fish, patent attorney and corporation president, assured his audience at the annual meeting of the AIEE in 1909. "In so far as there is any foundation for the contention that under modern conditions, the inventor himself does not get all that he should for his work, the basis for the contention is not the patent system or the law, but the social and industrial conditions which prevail."

CHAPTER 5

The New Industrial State

John Kenneth Galbraith

Endowed with a flair for writing with a sarcastic bite, John Kenneth Galbraith is a liberal economist who delights in deflating official myth and puncturing the conventional beliefs of his colleagues. His study of the American economy, *The New Industrial State*, attempts to show that it is the producers, not the consumers, that are truly sovereign in our economic system.

First published in 1967, the book has had an unusually long life. A second edition was brought out in 1971 and a third in 1978, with only minor changes deemed needed. This selection is taken from the most recent edition, which appeared in 1985, but you may be confused when you encounter the author's passing reference to the "eleven years since the first edition of this appeared," as if it were still 1978 when he was writing. Indeed it *was* 1978 when he wrote the selection here: For the 1985 edition he made no changes and merely added a new introduction. He explains, "There comes a time when revision of a book must be brought to a halt and error or obsolescence accepted as one of the inescapable tendencies of this art form."

The New Industrial State is not a formal history of the modern industrial system, but it does reach into the past to show changes over time, particularly in the past fifty years. In this selection, Galbraith's wit is trained on the common misunderstanding of the word *planning*, which in the American mind came to be associated with communistic danger during the Cold War. The author argues that right under our noses the modern American corporation began to practice planning in everything but name anyhow. Companies plan in the sense that they invest in long-term research for new product development, arrange for long-term sources of supply at set prices, and make certain that there will be sufficient future customers by securing government guarantees in advance and by other means. Planning is used to reduce future uncertainties.

Such planning helps corporations operate profitably, but Galbraith's central argument is not that planning is primarily economically motivated, but rather it is *technologically* motivated, required by today's skill- and capital-intensive technology. The author illustrates by examining road construction logistics. If the road is simple and the technology crude, unskilled ditchdiggers and their crude picks and shovels are readily available on short notice. But for construction of a modern superhighway, skilled labor and heavy construction equipment are required and will not necessarily be readily available. Planning arrangements for the necessary resources must be done in advance.

The modern corporation uses a number of tactics to deal with the risk that future markets will be unreliable or inhospitable. Market research allows them to learn about trends in current demands, and Galbraith notes parenthetically that learning about consumer interests easily merges into shaping consumer interests. Protection from future loss is also secured by achieving massive size and diversifying into many different industries. The large conglomerate need not worry unduly about a misstep in one or another of its many divisions.

Galbraith accords primary emphasis to three other tactics that help corporations reduce uncertainty. Vertical integration, in which a company controls more than one stage in the vertical chain of supply, production, and distribution, is now found in a number of industries (we might mention McDonald's, which owns its own cattle ranches as an example). Market uncertainty is also reduced, the author claims, by the large corporation's efforts to set prices as it wishes. Gargantuan size and unspoken price agreements among the dominant firms permit them to dictate both the prices paid to suppliers and the prices paid by customers. Control of the market's operation is consolidated further by long-term contracts with other major firms or with government agencies.

In industries based on "exacting technologies" (what journalists now call high tech), research and development for new products requires long lead time and massive investment, and it entails an extremely high risk of failure. The task is so difficult, Galbraith tells us, that only the very largest firms have the resources for such undertakings, and even then, government guarantees of one sort or another are often demanded.

Yet we can ask if Galbraith's argument holds true today. He has not essentially revised it since he first made it, in the 1960s. During the past ten years, however, we have seen thousands of small electronics- or biotechnology-related companies spring up in many parts of the country, Silicon Valley in California and Route 128 in Massachusetts being only the best-known concentrations. The new technologies they have developed or refined have been widely acknowledged as being consistently more advanced than those developed by larger and older rivals. After all, the small firm has no resource to draw upon other than technological leadership. It was Apple Computer, not IBM, that secured a reputation for innovative technology. The question of which kind of company—well-financed behemoth or hungry start-up—is most likely to produce future technological advances is one that occupies national policymakers today, for the answer determines where federal resources and tax incentives should be channeled.

THE NATURE OF INDUSTRIAL PLANNING

Until the end of World War II or shortly thereafter, planning was a moderately evocative word in the United States. It implied a sensible concern for what might happen in the future and a disposition, by forehanded action, to forestall avoidable disfunction or misfortune. As persons won credit for competent planning of their lives, so communities won credit for effective planning of their environment. It was thought good to live in a well-planned city. The United States government before the war had a National Resources Planning Board. During the war, postwar planning acquired the status of a modest industry in both the United States and the United Kingdom; nothing else, it was felt, would so reassure those who were fighting that they had eventual utility as civilians.

In the Cold War years, however, the word planning acquired grave ideological overtones. The Communist countries not only socialized property, which seemed not a strong likelihood in the United States, but they planned, which seemed more of a danger. Since liberty was there circumscribed, it followed that planning was something that the libertarian society should avoid. Modern liberalism carefully emphasizes tact rather than clarity of speech. Accordingly, it avoided the term, and conservatives made it one of opprobrium. For a public official to be called an economic planner was less serious than to be charged with Communism or imaginative sexual proclivity but it reflected adversely nonetheless. One accepted and cherished whatever eventuated from the untrammeled operation of the market. Not only concern for liberty but a reputation for economic hardihood counseled such a course.

For understanding the economy and polity of the United States and other advanced industrial countries, this reaction against the word planning could hardly have been worse timed. It occurred when the in-

SOURCE: From *The New Industrial State*, 4th ed., by John Kenneth Galbraith. Copyright © 1967, 1971, 1978, 1985 by John Kenneth Galbraith. Reprinted by permission of Houghton Mifflin Company.

creased use of technology and the accompanying commitment of time and capital were forcing extensive planning on all industrial communities—by firms and of firms' behavior by government. The ban on the use of the word planning excluded reflection on the reality of planning.

This ban is now in the process of being lifted—much has been accomplished in this regard in the eleven years since the first edition of this book appeared. The need for national planning has become a reputable topic for discussion, as also legislation to facilitate it. On a matter such as energy the need is accepted but in circles of the highest repute the term czar is still preferred to that of planner, though not, one judges, because it is deemed more democratic.

However, it is still the instinct of conservatives and those for whom high banking or corporate position serves as a substitute for thought that anything called planning should be resisted. And perhaps there are useful elements of self-interest in the effort. Any discussion of planning by the government will draw attention, inevitably, to the planning by corporations that makes it necessary. Those who now, in the manner of all planners, guide or control the behavior of individuals will no longer be able, on grounds of high principle, to resist public guidance, control or coordination of *their* planning.

In the market economy the price that is offered is counted upon to produce the result that is sought. Nothing more need be done. The consumer, by his offer to pay, obtains the necessary responding action from the firm that supplies his needs. By offering to pay yet more, he gets more. And the firm, in its turn, by similar offers gets the labor, materials and equipment that it requires for production.

Planning exists because this process has ceased to be reliable. Technology, with its companion commitment of time and capital, means that the needs of the consumer must be anticipated—by months or years. When the distant day arrives, the consumer's willingness to buy may well be lacking. By the same token, while common labor and carbon steel will be forthcoming in response to a promise to pay, the specialized skills and arcane materials required by advanced technology cannot similarly be counted upon. The needed action in both instances is evident: in addition to deciding what the consumer will want and will pay, the firm must take every feasible step to see that what it decides to produce is wanted by the consumer at a remunerative price. And it must see that the labor, materials and equipment that it needs will be available at a cost consistent with the price it will receive. It must exercise control over what is sold. It must exercise control over what is supplied. It must replace the market with planning.

That, as more time elapses and more capital is committed, it will be increasingly risky to rely on the untutored responses of the consumer needs no elaboration. And this will be increasingly so the more technically sophisticated the product. There is a certain likelihood that even two or three years hence there will be a fairly reliable consumer demand for strawberries, milk and fresh eggs. There is no similar assurance that people will want, so spontaneously, an automobile of particular color or contour or a transistor of particular size or design.

The effect of technology and related change in reducing the reliability of the market for labor or equipment and in making imperative the planning of their procurement is equally clear and can be seen in the simplest case. If men use picks and shovels to build a road, they can be called out on the same morning that the decision is taken to do the job. The picks and shovels serve a variety of purposes; accordingly, the market stocks them in readily available quantities. It will help in getting manpower if, as Marx thought necessary, there is an industrial reserve army of the unemployed. But an equally prompt beginning is possible by raiding the work force of another employer of unskilled labor with the simple market promise of more pay.

When specifications are raised to modern superhighway standards and heavy machinery is introduced, the market no longer works as well. Engineers, draftsmen, drainage experts and those who arrange the elimination of trees, grass, parkland, streams and the other environmental amenities may not be readily available even in response to a substantial advance in pay. Bulldozers and heavy earth-moving equipment cannot be bought with the same facility as picks and shovels. In all of these cases anticipatory steps must be taken to ensure that the necessary supply is available at an appropriate wage or price. Market behavior must be modified by some measure of planning.

For inertial systems engineers, digital circuit design specialists, superconductivity research specialists, aeroelasticity investigators and radio test and evaluation engineers, as also for titanium alloys in comparison with steel and space vehicles as compared with motorcycles, the market is greatly less dependable. Need must be elaborately anticipated and arranged. The language of both industry and government reflects the modern fact. Civil War quartermasters went into the local markets for their needs. So, in turn, did the contractors who filled these orders. The equivalent procurement would now have to be programmed months or years ahead.

As viewed by the industrial firm, planning consists in foreseeing the actions required between the initiation of production and its completion and preparing for the accomplishment of these actions. And it consists also of foreseeing, and having a design for meeting, any unscheduled developments, favorable or otherwise, that may occur along the way. As planning is viewed by the economist, political scientist or professional sage, it consists of replacing prices and the market as the mechanism for determining what will be produced with an authoritative determination by the state of what will be produced and consumed and at what price. It will be thought that the word planning is being used in two different senses. Planning by the firm—its long-range accommodation to the market influences to which it is subject—is surely different from exterior planning which stipulates what its prices and production will be.

In practice the two kinds of planning, if such they may be called, are inextricably associated. A firm cannot satisfactorily foresee and schedule future action or prepare for contingencies if it does not know what its prices will be, what its sales will be, what its costs, including labor and capital costs, will be and what will be available at these costs. If the market is uncontrolled, it will not know these things. If, with advancing technology and associated specialization, the market becomes increasingly unreliable, as we have seen it does, industrial planning will become increasingly difficult unless the market is also controlled, made part of the planning. Much of what the firm regards as planning consists in minimizing uncontrolled market influences.

A variety of strategies are available for dealing with the increasing unreliability of markets. If the item is unimportant, market uncertainty can be ignored. For General Electric it is a matter of considerable interest to know the price at which it will be able to buy high-alloy steel or sell large generators and the quantities that will be forthcoming or that can be sold. No similar urgency attaches to knowledge of the price at which flatware will be available for the plant cafeterias. Something, and perhaps much, can also be learned about the prospective market behavior of consumers by market research and market testing. (Research into what the consumer wants or will want merges into research in how the consumer can best be persuaded.) And, finally, large-scale operations allow some market uncertainty to be absorbed. In 1977, one of the three big Swiss banks, Crédit Suisse, in consequence of what must have been a uniquely massive gap in its management controls, failed to monitor some exceptionally imaginative operations of its Chiasso branch on the Italian border. Losses of several hundred million dollars resulted. So great was the scale of banking operations that these could be absorbed, though not without some promise of improved managerial performance. In the same period, through similar but less inspired incompetence, the large New York banks, Chase Manhattan in particular, had huge losses on speculative real estate and ill-considered foreign loans. These also were absorbed by size, although again not without some need to promise more enlightened future performance. Earlier, in the late nineteen-fifties and early nineteen-sixties, the Convair Division of General Dynamics Corporation lost $425 million on the manufacture of jet transports. Part of this was the result of uncertainties associated with research and development; its 880 and 990 passenger jets cost more to bring into being than expected. But a major factor was the failure of the market—or, more precisely, default on or failure to obtain the contracts that were meant to reduce market uncertainty. The company did not fail (although it was a near thing) because it had annual revenues of around $2 billion from—in addition to aircraft—such diverse artifacts as missiles, building materials, submarines and telephones. None of these was affected by the misfortunes of Convair. For a smaller company, with one product, a $425 million loss would have been uncomfortable. We have here a partial explanation of the origins of one of the more notable corporate developments of recent times, the growth of the conglomerate corporation. It combines great size with highly diverse lines of manufacture. Thus it can absorb the adverse consequences of uncertainty that cannot otherwise be eliminated. Uncontrolled aversion of customers to one product, such as aircraft, is unlikely to affect telephones or building materials. The effects of market uncertainty are thus contained in what will often be a relatively small part of the total planning unit.

But the more common strategies require that the market be replaced by an authoritative determination of price and the amounts to be sold or bought at these prices. There are three ways of doing this:

1. The market can be superseded.
2. It can be controlled by sellers or buyers.
3. It can be suspended for definite or indefinite periods by contracts between the parties to sale and purchase.

All of these strategies are familiar features of the planning system.

The market is superseded by vertical integration. The planning unit takes over the source of supply or the outlet; a transaction that is subject to bargaining over prices and amounts is thus replaced with a transfer within the planning unit. Where a firm is especially dependent on an important material or product—as an oil company on crude petroleum, a steel firm on ore, an aluminum company on bauxite or Sears, Roebuck on appliances—there is always danger that the requisite supplies will be available only at inconvenient or uncertain prices. To have control of supply—to rely not on the market but on its own sources of supply— is an elementary safeguard. This does not eliminate market uncertainty; rather the large and unmanageable uncertainty as to the price of ore or crude is replaced by the smaller, more diffuse and more manageable uncertainties as to the costs of labor, drilling, ore transport and yet more remote raw materials. But this is a highly beneficial exchange. For an oil company a change in the cost of purchased crude is a serious matter, a change in the cost of drilling equipment a detail.

As viewed by the firm, elimination of a market converts an external negotiation and hence a partially or wholly uncontrollable decision to a matter for purely internal decision. Nothing, we shall see, better explains modern industrial policy—capital supply is the extreme case—than the desire to make highly strategic cost factors subject to wholly internal decision.

Markets can also be controlled. This consists in reducing or eliminating the independence of action of those to whom the planning unit sells or from whom it buys. Their behavior being subject to control, uncertainty as to that behavior is reduced. At the same time the outward form of the market, including the process of buying and selling, remains formally intact.

This control of markets is the counterpart of large size and large size in relation to the particular market. A Wisconsin dairy farm cannot influence the price that it pays for fertilizer or machinery. Its purchases being small in relation to purchases as a whole, its decision to buy or not to buy has no appreciable significance for the supplier. The same is true of its sales. Having no control over its suppliers or its customers, it pays and receives the going prices.

Not so with General Motors. Its decision to buy or not to buy will usually be very important to its suppliers; it may be a matter of survival. This induces a highly cooperative posture. So with any large firm. Should it be necessary to press matters, General Motors, unlike the dairyman, has always the possibility of supplying a material or component to itself. The option of eliminating a market is an important source of power for controlling it.

Similarly, size allows General Motors as a seller to set prices for automobiles, diesels, trucks, refrigerators and the rest of its offering and be secure in the knowledge that no individual buyer, by withdrawing his custom, can force a change. The fact that GM is one of a few sellers adds to its control. Each seller shares the common interest in secure and certain prices; it is to the advantage of none to disrupt this mutual security system. Competitors of General Motors are especially unlikely to initiate price reductions that might provoke further and retributive price-cutting. No formal communication is necessary to prevent such actions; this is considered naïve and arouses the professional wrath of company counsel. Everyone knows that the survivor of such a contest would not be the aggressor but General Motors. Thus do size and small numbers of competitors lead to market regulation.

Control of prices is not a part of market control; if uncertainty is to be eliminated, there must also be control of the amount sold. But size also makes this possible. It allows advertising, a well-nurtured sales organization and a careful management of product design which can help to ensure the needed customer response. And since General Motors produces some half of all the automobiles available from domestic sources, its designs do not reflect the current mode but are the current mode. The proper shape of an automobile, for most people, is what the automobile majors decree the current shape to be. The control of demand, as we shall see later, is not perfect. But what is imperfect is not unimportant for reducing market uncertainty.

Finally, in an economy where units are large, firms can eliminate market uncertainty for each other. This they do by entering into contracts specifying prices and amounts to be provided or bought for substantial periods of time. A long-term contract by the Wisconsin farmer to buy fertilizer or sell milk accords no great certainty to the fertilizer dealer or the dairy receiving the milk. It is subject to the capacity of the farmer to fulfill it; death, accident, drought, high feed costs and contagious abortion can all supervene. But a contract with the United States Steel Corporation to supply sheet steel or to take electric power is extremely reliable. In a world of large firms, it follows, there can be a matrix of contracts by which each firm eliminates market uncertainty for other firms and, in turn, gives to them some of its own uncertainty.

Outside of the planning system, most notably in agriculture, the government also intervenes extensively to set prices and ensure demand and thus to suspend the operation of the market and eliminate market uncertainty. This it does because the participating units—the individual farms—are not large enough to control prices. Technology and the associated commitment of capital and time require, nonetheless, that there be stable prices and assured demand. But within the planning system similar action is also required where exacting technology, with extensive research and development, means a lengthy production period and a very large commitment of capital. Such has long been the case in the development and supply of modern weapons; it was and remains true in the exploration of space; and it characterizes the development of a growing range of modern civilian products or services, including transport planes, high-speed ground transport, various applied uses of nuclear energy and sundry new sources of energy or forms of energy conservation. Here the state guarantees a price sufficient, with suitable margin, to cover costs. And it undertakes to buy what is produced or to compensate fully in the case of contract cancellation, technical failure or absence of demand. Thus, effectively, it suspends the market with all its associated uncertainty. One consequence, as we shall see, is that, in areas of the most exacting advanced technology, the market is most completely replaced, and planning is therefore most secure. As a further consequence, this has become for the participants a very attractive part of the planning system. The fully planned economy, so far from being unpopular with avowed friends of free enterprise, is warmly regarded by those who know it best.

Two things of some interest are evident from this analysis. It is clear, to repeat, that industrial planning is in unabashed alliance with size. The large organization can tolerate market uncertainty as a smaller firm cannot. It can contract out of it as the smaller firm cannot. Vertical integration, the control of prices and consumer demand and reciprocal absorption of market uncertainty by long-term contracts between firms all favor the large enterprise. And while smaller firms can appeal to the state to fix prices and ensure demand, this security is also provided by the state to the big industrial firm when it is most needed. These circumstances—the exacting technology, large commitments of time and capital—make it fairly certain that most of this government work will be done by large organizations.

By all but the pathologically romantic, it is now recognized that this is not the age of the small man. But there is still a lingering presumption among economists that his retreat is not before the efficiency of the great corporation or even its technological proficiency but before its monopoly power. The corporation has superior capacity to extract profits. Therein lies its advantage. "Big business will undertake only such innovations as promise to enhance its profits and power, or protect its market position. . . . [F]ree competitive men have always been the true innovators. Under the stern discipline of competition they must innovate to prosper and to survive."

This, by the uncouth, would be called drivel. Size is the general servant of technology, not the special servant of profits. The small competitive firm cannot afford the outlays that innovation demands. An economic system consisting of such firms would require, rather, that we reject the technology which, since earliest consciousness, we have been taught to applaud. It would require that we have simple products made with simple equipment from readily available materials by unspecialized labor. Then the period of production would be short, the market would reliably provide the labor, equipment and materials required for production; there would be neither possibility nor need for managing the market for the finished product. If the market thus reigned, there would be, and could be, no planning. The small firm would then, at last, do very well. All that would be necessary would be to undo nearly everything that, at whatever violence to meaning, has been called progress in the last half century. There may be a case against technical innovation—against supersonic travel and infinitely destructive weaponry and even different automobiles and detergents. There is no case that such innovation will come better from the little man.

The second conclusion is that the enemy of the market is not ideology but the engineer. In the Soviet Union and the Soviet-type economies prices are extensively managed by the state. Production is not in response to market demand but given by the overall plan. In the Western economies markets are dominated by great firms. These establish prices and seek to ensure a demand for what they have to sell. The enemies of the market are thus highly visible, although rarely in social matters has there been such a case of mistaken identity. They are not socialists. The enemies, in both cases, are advanced technology, the specialization and organization of men and process that this requires and the resulting commitment of time and capital. These make the market work badly when the need is for greatly enhanced reliability—when planning is essential. The modern large Western corporation and the modern apparatus of socialist planning are variant accommodations to the same need. It is open to every free-born man to dislike this accommodation. But he must direct his attack to the cause. He must not ask that jet aircraft, nuclear power plants or even the modern automobile in its modern volume be produced by firms that are subject to unfixed prices and unmanaged demand. He must ask, as just noted, that they not be produced.

CHAPTER 6

Birth Control in America: The Career of Margaret Sanger

David M. Kennedy

The life of Margaret Sanger (1883–1966) spanned two separate eras in American medical and social history. When Sanger, trained as a nurse, began her crusade to make birth control knowledge, and simple, inexpensive techniques available to all who wished to use them, she faced seemingly invincible opposition from all the major institutions in American society. Moreover, the most commonly used contraceptives, even as late as 1943, when she retired from active leadership in the birth control movement, were still intravaginal devices, chemicals, and condoms. But by the 1960s, when she died, oral contraceptives were taken by millions of women around the world, and many of the institutions that formerly had opposed birth control now supported it. Presidents who had denounced birth control now publicly embraced family planning. This remarkable transformation was not the simple technical result of laboratory innovation that created the Pill. Rather it was the result of long-fought social, legal, and political battles, led by Sanger, battles that preceded, and brought into being, the laboratory that eventually produced the Pill. David Kennedy's book tells the story of these earlier battles in the first decades of the twentieth century, and of Margaret Sanger, who was the tireless birth control advocate on the front line (and who is credited with coining the very term *birth control*).

Our selection, "Birth Control and American Medicine," examines the stubborn resistance of the medical profession to acceptance of birth control measures, up to as late as the 1930s. Between 1910 and 1920, the period Kennedy discusses first, Sanger's own thinking about the medical profession changed in a significant way. Until 1915, Sanger had thought that women, once shown by a nurse or physician how to properly use a diaphragm, could teach one another, with no further need of consulting the medical profession. But in early 1915, she went to Holland to visit a government-supported birth control clinic, where she met the clinic's director, Dr. Johannes Rutgers. He convinced her that every woman who desired birth control instruction should be medically examined and diaphragms professionally fitted. Convinced by Rutgers that contraception was a medical matter, Sanger returned to the United States and sought support from the medical profession.

One more bit of background information may be helpful. The New York State Penal Code made the distribution of birth control devices all but legally impossible. Section 1142 made it a misdemeanor for anyone to "sell, lend, or give away," or to advertise, loan, or distribute, "any recipe, drug or medicine for the prevention of conception." Section 1145 provided a limited exception, however, by sanctioning any "article or instrument" that physicians used "for the cure or prevention of disease." But no physicians were willing to work in the birth control clinic—the nation's first—that Sanger set up in 1916 in the Brownsville section of Brooklyn. She was arrested for "maintaining a public nuisance" and in 1917 served thirty days in the workhouse.

Physicians had a number of reasons why they would not sanction birth control. Kennedy shows that the rank and file of American doctors believed that birth control measures were medically unsafe and morally destructive. If birth control was truly necessary, doctors prescribed continence (simply abstaining from sex) or sterilization.

Kennedy also cites as a cause of physician resistance the medical profession's new sense of professionalism and its opposition to quacks— the medical faddists, bleeders, and patent-medicine salesmen—who had so recently preyed upon the American public in distressingly large numbers. When Margaret Sanger claimed that birth control would end poverty, disease, and virtually all evils in the world, American doctors recoiled, hearing echoes of the extravagant claims of the patent-medicine salesmen.

Convincing physicians of the legitimacy of prescribing birth control measures for strictly medical reasons (or in the technical parlance that Kennedy uses, for medical "indications") proved to be much easier than for nonmedical reasons. Sanger urged physicians to take on responsibility not only for medical problems but also for economic and social problems, and this they were most unwilling to do.

Although lacking the sanction of their doctors, thousands of women in the 1920s went ahead and prescribed birth control devices, such as the cervical cap, for themselves. The devices were crude and often led to infection, but most physicians only renewed their condemnation. Lonely and exceptional was the call of a doctor like Robert Latou Dickinson, who proposed new research to investigate safer, more effective techniques.

In the 1920s, Sanger had to literally smuggle in contraband shipments of diaphragms, but after 1930, contraceptives could be shipped legally "for the prevention of disease." Advertisements appeared, sales grew exponentially, and still the American Medical Association hesitated to adopt a resolution urging that state and federal laws be amended so that physicians could legally provide contraceptive information to their patients. In 1937, after two years of committee meetings, changes in membership and changes in positions, an AMA committee accepted the position that Margaret Sanger had promoted for two decades—that birth control measures could be prescribed for nonmedical reasons, including the simple one that the patient desired it. This stance required a revolution in the thinking of the medical profession.

One of the most controversial aspects of the contraception debate—controversial now as well as then—is the eugenic argument for birth control, that is, controlling the births of certain groups to "improve" the general population. Eugenics has been proposed by advocates of many goals—from raising the average intelligence of a country to eliminating nonwhite groups from the population. Sanger herself was a eugenicist—elsewhere she wrote that birth control was supposed to help weed out the "unfit," which she defined broadly as the "feebleminded," but which others defined in explicitly racial, ethnic, or class terms.

Look for signs of eugenic thinking as you read the selection; they surfaced here and there between 1900 and 1930. Today, concern about eugenics has replaced the earlier medical opposition as one of the primary impediments to birth control information. Some members of minorities in the United States and citizens of many nonwhite countries in the Third World are extremely suspicious of United States- or United Nations-sponsored family planning programs, seeing in them a disguised form of eugenics being practiced against nonwhite peoples. Planned Parenthood clinics in minority neighborhoods have been attacked for allegedly attempting to carry out racial genocide.

Continuing moral and religious objections to so-called artificial birth control also block its extension. Technically speaking, unquestionable advances since the early twentieth century have been made, but birth control technology remains inextricably bound to the social setting, and acceptance remains far from universal.

BIRTH CONTROL AND AMERICAN MEDICINE

At Dr. Johannes Rutgers's clinic in 1915, Margaret Sanger had learned the indispensability of medical support to the success of birth control. Only doctors could ensure a safe and effective contraceptive technique. The New York Court of Appeals' decision in 1918 added, at least for New York State, another imperative: only licensed physicians could prescribe birth control legally. Accordingly, Mrs. Sanger for the next two decades pleaded for medical endorsement of contraception. During that time the tortured relationship between Margaret Sanger and the organized medical profession revealed much about the birth control movement and about the nature of American medicine.

Mrs. Sanger had made scant reference to medical doctors in her pamphlet "Family Limitation" and none at all in the Woman Rebel. Others, however, had considered contraception an essentially medical question well before Mrs. Sanger learned that lesson in Holland. In the 1830s a Massachusetts physician, Charles Knowlton, had written the most famous of the nineteenth-century "underground" tracts on contraception. In the 1880s and 1890s, medical doctors not infrequently discussed contraceptive techniques in their professional journals. Rarely, however, did they mention the subject publicly. Then, in his presidential address to the American Medical Association in 1912, the respected pediatrician Abraham Jacobi cited the high fertility of immigrants, the rising costs of welfare, and the burdens of poverty to substantiate his "indispensable suggestion that only a certain number of babies should be born into the world." Jacobi had been influenced by his friend William J. Robinson, an eccentric physician with his own medical publishing business. Robinson had long believed in contraception, or what he called "prevenception," and had advocated though never actually described it in his various journals and books.

Robinson labored alone for years, with Jacobi his only important convert. But in May 1915, noting the public interest in Margaret Sanger's flight to Europe and William Sanger's arrest, Robinson and the feminist leader Henrietta Rodman arranged an informal discussion of birth control at the New York Academy of Medicine. Though little came of the session, it represented a medical initiative at least indirectly in response to Mrs. Sanger's activities. Most of the discussants came to the meeting already sympathetic toward birth control; four of them—Robinson, Lydia Allen DeVilbiss, Ira S. Wile, and S. Adolphus Knopf—had for some time advocated greater medical participation in the investigation and practice of contraception. Wile and especially Knopf, a tuberculosis specialist, enjoyed substantial prestige among their fellow physicians. Similarly, a meeting in Chicago in February 1916 presented reputable doctors like Alice Hamilton and Rachelle S. Yarros of Hull House who spoke in favor of birth control. A distinguished guest from New York, Robert Latou Dickinson, one of the country's most eminent gynecologists,

SOURCE: From Birth Control in America, by David M. Kennedy. Copyright © 1970 by Yale University. Reprinted by permission of Yale University Press.

urged "that we as a profession should take hold of this matter [of contraception] and not let it go to the radicals, and not let it receive harm by being pushed in any undignified or improper manner."

The rank and file of American medicine did not share that opinion. The associations, academies, and societies which alone could make an innovation such as birth control acceptable to the average doctor refused to endorse it. In late 1916, the New York County Medical Society took the first formal action on birth control of any organized American medical body. A Medical Society committee, studying a proposed amendment to New York's Section 1142, reported its fear that contraceptives, "indiscriminately employed," would undermine personal morality and national strength. The committee suspected that professional abortionists and "sensation-mongers" were behind much of the birth control propaganda. A minority report argued that tuberculosis and epilepsy should be considered legitimate medical indications for contraception and that birth control would end wars, child labor, drunkenness, divorce, poverty, and crime. But the majority countered that Section 1145 already made contraception legitimate for "good and sufficient reasons" (such as serious illness), and renounced any connection between birth control, as a purely medical matter, and social and economic problems. Finally, the committee pointed to medical ignorance of a reliable, harmless contraceptive method. Many "foreign procedures," it insisted, were "absurd, frequently dangerous, filthy, and usually unsatisfactory." It recommended that the Medical Society repudiate "ill-advised legislative enactments and sensational propagandistic movements."

Medical men generally favored the views of the committee's majority. In an address to the New York Obstetrical Society in January 1917, George Kosmak, a prominent gynecologist, later editor of the American Journal of Gynecology, asked:

> Shall we listen to the unrestrained harangue of the reformer, usually a lay person with little conception of the medical aspects involved, who within the narrow bounds of his or her vision sees the solution of all the faults of our social system eliminated through the dissemination of contraceptive measures? Shall we lend ourselves to the spirit of license which such sentiments naturally must convey?

Kosmak dismissed the birth control advocates as "radical socialists" and "anarchists," whose unscientific literature on contraception contained "arrant nonsense, false reports, and . . . seditious libels on the medical profession." But he did encourage physicians to recommend continence—the only sure method, he said, whatever its liabilities—in cases where pregnancy would endanger a woman's life.

Another writer allowed that birth control was indicated "in all chronic diseases"; beyond that, "economics and medicine are at the parting of the ways." But since there was absolutely no effective contraceptive, and since "the female is harmed by all devices which prevent the entrance of sperma [sic] in the uterine cavity," the only acceptable contraceptive—even in those cases where it was essential to health—was sterilization.

For all their objections to birth control, most doctors clearly did not reject the practice outright. They usually gave guarded approval to contraception in cases with pathologic indications (conditions warranting therapy), but in general they rejected artificial devices in favor of continence or sterilization as the best prophylaxis. That severely qualified approval fell far short of Margaret Sanger's hopes for public support.

At first Mrs. Sanger ascribed medical reticence to moral cowardice and the prejudices of bourgeois doctors insensitive to lower-class suffering and opposed to social justice. To be sure, taking the Hippocratic oath did not automatically liberate a doctor from the values of his place, class, and time. The attitude of most American physicians toward the early birth control movement unmistakably reflected their middle-class upbringing and their commitment to nineteenth-century sexual mores. Yet early medical reaction to birth control derived principally neither from simple cowardice nor from strictly class bias; it owed most, rather, to a relatively new sense of medical professionalism. In commenting on contraception, medical men struck three persistent notes. Doctors had a reflex aversion to anything that smacked of lay medicine, sensationalism, or quackery. They opposed other than pathologic indications for medical treatment. Most forcefully, they rejected any therapeutic technique so untried as artificial contraception. All these themes proceeded from the history of medicine in nineteenth-century America. . . .

Robert Latou Dickinson played on the deep-seated fear of quackery among medical men when in 1916 he urged his colleagues who supported birth control to keep it out of the hands of the "radicals." That Margaret Sanger was a radical in the conventional, political sense amplified the predictable medical panic at her paramedical activities. George Kosmak and other doctors heard echoes of the patent-medicine barker when Mrs. Sanger claimed that birth control would end poverty, disease, crime, war, and all manner of social evils. Birth control propaganda sounded remarkably like the most extravagant quack advertising against which orthodox medicine had long battled. Most physicians, therefore, opposed Mrs. Sanger as a dangerous, unorthodox interloper. Even those sympathetic to her cause, like Dickinson, sought to reduce or eliminate her personal influence in order to make birth control acceptable to the profession at large.

Doctors who favored contraception also found that they alienated their colleagues when they discussed nonmedical indications. When Abraham Jacobi spoke of immigration, welfare, and poverty in relation to family restriction, he went beyond the dimensions of contraception as a purely medical matter—dimensions within which his fellow physicians would much more willingly have responded. Medical leaders in contraception soon realized the resistance of the profession to "social" indications and confined their discussion to the necessity of birth control in well-defined pathologic circumstances—tuberculosis, heart or kidney disease, and pelvic abnormality.

All advocates of birth control recognized that the major source of medical opposition to the practice lay in ignorance of an acceptable contraceptive technique. Dickinson, in his presidential address to the American Gynecological Society in 1920, noted that ignorance—and ignorance of sex in general—and demanded a systematic effort to overcome it. "What serious study," he asked,

> has ever been made bearing upon the harm or harmlessness of the variety of procedures or concerning the failure or effectiveness of each? Who has or can acquire any considerable body of evidence on these matters but ourselves? What, indeed, is normal sex life? What constitutes excess or what is the penalty for repression in the married? Do we still have to hark back to Luther for an answer? It will take a few professional life-times of accredited histories to gather evidence to submit, but some time a start must be made.

Mrs. Sanger's move toward respectability during and immediately after World War I helped to allay some medical fears. She stopped carping at bourgeois doctors as she delivered the movement into the arms of middle-class ladies. And she began to acknowledge, reluctantly, her inadequacies as a laywoman and the consequent need for professional involvement. She would retire from the field, she hinted, as soon as doctors stepped in and fulfilled their responsibilities in the matter of contraception.

On the question of indications, however, Mrs. Sanger stood her ground. The unfavorable report of the New York County Medical Society, said the *Birth Control Review*, "furnished the final proof that birth control is not a medical but a social question . . . the economic side is much more important than the purely pathologic." After Judge Crane's 1918 decision, Mrs. Sanger urged New York physicians to give the widest possible construction to Crane's liberal definition of "disease." Medicine, she said, should broaden its conception of its responsibilities to include the amelioration of eugenic, economic, and social problems through the application of medical knowledge.

Mrs. Sanger also squarely faced the doctors' demand for more knowledge of the technical aspects of contraception. At the First American Birth Control Conference in 1921 she announced her intention to open a clinic. The clinic, she said, would of course provide service, as the Brownsville clinic had done. But there the resemblance stopped. Unlike Brownsville, designed primarily as a legal test and publicity stunt, the new clinic would above all be a "first-class" center for medically supervised study of contraceptive techniques. The results of the study, Mrs. Sanger believed, would prove to the world and to the medical profession in particular the desirability and efficacy of birth control. But the doctor who was to be head of the clinic, Lydia Allen DeVilbiss, refused to go ahead until Mrs. Sanger received a proper dispensary license from the State Board of Charities. The board did not grant the license, Dr. DeVilbiss went off to practice in Florida, and the clinic did not open.

Characteristically, Mrs. Sanger continued her efforts to open a center for contraceptive service and study. By late 1922, she had hired Dr. Dorothy Bocker, a public health officer from Georgia, who "knew practically nothing about birth control technique, but was willing to learn." Dr. Bocker had to operate the "clinic" as her private practice. In compliance with medical ethics, she could not advertise, and in compliance with state law she could prescribe only for a "health reason." Mrs. Sanger leased offices for the clinic across the corridor from the American Birth Control League headquarters at 104 Fifth Avenue. The ABCL, legally able to advocate birth control, but not to prescribe it, could simply refer inquiries to Dr. Bocker's "Clinical Research Bureau" across the hall. It was an arrangement of dubious propriety, conforming to the letter but hardly to the spirit of medical ethics and the New York statutes. Mrs. Sanger warned Dr. Bocker that she might go to jail for her efforts, but offered some consolation: "If this does happen, I believe you will get such a good boost of publicity, that we can put you on the platform lecturing throughout the country for the next two years." Dr. Bocker, hoping to escape harassment long enough to build a body of convincing statistical information on birth control, opened the doors of the Clinical Research Bureau on January 2, 1923.

The bureau, expecting to reach primarily working-class women, offered its services free. But because of the lack of advertising and the nature of the ABCL's membership, the "educated classes" were "unduly represented" among the 1,208 patients who came to the clinic in the first twelve months of its operation.

Dr. Bocker experimented with various contraceptive techniques. The most successful proved to be the combination of a spermacidal jelly with a Mensinga type diaphragm. Mrs. Sanger had procured the basic jelly formula in Germany in 1920, and it was soon being manufactured in the United States. American manufacturers, however, produced only the unsatisfactory "check," or cervical-cap diaphragm, and Federal statutes made it illegal to import contraceptive supplies. Hence the few Mensinga diaphragms the clinic received came from Germany to Mr.

Slee's Three-in-One Oil plant in Montreal. From there Slee ran the contraband diaphragms across the Canadian border in oil drums. But even that elaborate operation could not keep the clinic adequately supplied, and Mrs. Sanger therefore encouraged the development of a domestic manufacturing concern to produce a quality Mensinga diaphragm. The Holland-Rantos Company took form in 1925 in response to her request. Mrs. Sanger, to ward off the charge that she was in the birth control movement for profit, scrupulously refrained from any connection with that necessarily commercial enterprise.

The clinic went unmolested and Dr. Bocker continued through 1923 and 1924 to gather information on various contraceptive techniques. She published a report on over a thousand cases in 1924, but since none of her thirteen test series included more than a hundred women, the results were of limited persuasiveness. But the clinic's very existence, and the growing popular response to Mrs. Sanger's propaganda, began to command the interest of doctors.

One of the most interested physicians in the 1920s was the prominent gynecologist Robert Latou Dickinson. His career as a teacher and lecturer made his triangular face, accentuated by a trim Van Dyke beard, familiar throughout the medical profession. The public knew him as a nature writer and authority on New Jersey's Hudson palisades. His early concern with the pathologic effects of women's dress and traditional avoidance of exercise, and later his sympathetic opinion on birth control, marked him as a progressive ally of the feminist movement. His colleagues honored him in 1920 with the presidency of the American Gynecological Society.

Dickinson knew that the greatest obstacle to medical support for contraception was ignorance of an acceptable technique. The doctors who attended the special medical section of the First American Birth Control Conference in 1921 spent the session warning each other about untested techniques. They doubted the efficacy and especially the safety of nearly every birth control method suggested. Three years later, Dickinson reported: "A guaranteed technique of contraception is not yet worked out." Though the public, in the early 1920s, seemed to be resorting increasingly to *coitus interruptus,* suppositories, intravaginal and intrauterine pessaries, and condoms, medical men generally doubted the propriety of all those devices. Since the meager scientific data on contraception allowed no conclusions as to the effectiveness of any known technique, most doctors, in the nineteenth-century tradition of negative therapeutics, refused to prescribe anything. Abstinence, believed the orthodox, was the only certain contraceptive.

Some doctors, however, did prescribe birth control; and thousands of women prescribed for themselves without ever consulting a medical authority other than the corner druggist or Margaret Sanger's writings.

Against those loose practices the profession waged incessant battle. Medical literature throughout the 1920s abounded in warnings of the harmful effects of contraceptive devices. Often doctors warned of psychological damage, but they chiefly cited hard evidence of physical harm. Writers frequently reported that many commercially advertised douching solutions and suppositories injured the vagina. As for pessaries, Dr. Wilhelm Mensinga had originally advised that his Mensinga diaphragm be left in place between menses, and many doctors and laywomen took Mensinga at his word. The unsanitary conditions that such a practice encouraged led to several reported cases of streptococcus infections and even, some thought, carcinoma of the cervix. Many believed that the damage done by intravaginal devices could lead to permanent sterility.

One device doctors universally rejected: the so-called wishbone, or stem, or gold-pin, cervical cap. Though few if any doctors prescribed it, the high incidence of its condemnation in the medical literature indicated that a great number of women used the "wishbone," with frequently disastrous results. Because this device did not simply cover the cervix, but also extended into the uterine canal, medical authorities hesitated to say exactly how it was supposed to stop conception. If it simply retarded the entry of spermatozoa into the cervix, it was a true contraceptive. If the stem's presence in the uterus prevented nidation of the zygote, then it was in fact an abortifacient, which posed ethical and medical problems of a quite different order. In most cases, however, the "pin" simply failed to work at all. An English doctor who had experimented with it said: "The cervix is kept patent by it, and the way is thus left open for the entry of septic organisms from the exterior which may reach the uterus and give rise to pathological conditions. It is not a reliable contraceptive; it often acts as an abortifacient; it has given rise to endo-cervicitis . . . and even death. It is a very dangerous instrument."

No doctor in the 1920s, however sympathetic to birth control, could ignore the medical evidence: there was no reliable contraceptive device; moreover, the increasingly popular "mechanical devices applied to the cervix" seemed "very dangerous" to more than one investigator. Dickinson, reviewing medical knowledge of contraception in 1924, warned: "The largest gap in the clinical reports is on the matter of injury or harm—local, or general, or to the nervous system or morals as the result of the use of any or all of these measures." He especially noted Mensinga's mischievous and unnecessary advice to leave the diaphragm in place for weeks. The medical literature on contraception, concluded Dickinson, was a "library of argument." With regard to efficacy, it was based on hearsay. With regard to safety, the scant evidence seemed to justify medical caution.

Dickinson asked, therefore, for a new program of contraceptive research, designed to find a technique both safe and effective. And to those two requirements he added a third: any effective contraceptive would have to be acceptable to the people intending to use it. To be adopted by the uneducated poor in particular, it must be aesthetically inoffensive, simple, and cheap. "Let us note," he said, "that the sheath and the douche appear to work poorly in the tenement. . . . Indeed, all measures show poorer results in the less intelligent." Even the combined diaphragm and jelly, which the Clinical Research Bureau had tested with some success, chiefly among "educated" women, required the use of a bathroom, and therefore would not work well "in the tenement." Dickinson cautioned researchers that they must take into consideration "the inevitable lack of privacy" when they looked for a means "specially adapted . . . to the poor." Doctors repeated that caveat throughout the next two decades, as they increasingly realized that though the diaphragm and jelly technique was tolerably suitable for many women, it was "neither simple nor certain enough to meet the demands of a very large body of women whom it is most desirable to restrain from propagating." In addition to its other drawbacks, it required individual examination and fitting, the kind of medical attention the poor rarely received. When Mrs. Sanger and others tried in the 1930s to bring birth control to rural Southern Negroes and the masses of the Orient, the pressure for a cheap, simple contraceptive increased. Medical technology had to take on a social dimension. . . .

Before 1930, contraceptives could neither be advertised nor sent through the mails legally. A Federal court decision in that year, however, allowed advertisement and shipment of contraceptive devices intended for legal use—in most states, "for the prevention of disease." Under cover of that and similar euphemisms such as "feminine hygiene," a booming business in contraceptives developed rapidly. Scores of unscrupulous profit-seekers swamped the field that Mrs. Sanger had persuaded a single reluctant firm to enter in 1925. One investigator estimated expenditures for contraceptive advertising in 1932 and 1933 at $935,000. Advertisements appeared in newspapers, magazines, and even in the mail-order catalogs of Sears, Roebuck and Montgomery Ward. One unethical manufacturer adopted the name "Marguerite Sanger Company" and moved his headquarters regularly from state to state to avoid prosecution. *Fortune* magazine reported in 1938 that American women spent over $210,000,000 annually for contraceptive materials and that "the medically approved portion of business in female contraceptives is pitifully small . . . as a result . . . millions of women have been duped and thousands of secret tragedies have been enacted." "Contraceptive information today is all under-cover work and devices are being sold at retail in violation of the law so that the entire matter is in about the same position as the Volstead Act," wrote one doctor. The *New*

Republic laid the blame for these conditions "on those members of the medical profession who have sidestepped the issue." Although the Federal Trade Commission stopped some of the worst practices, "neither the government, the A.M.A., nor any other organization will give [a woman] any advice as to the relative merits of these products."

George Kosmak and many other doctors had condemned Margaret Sanger in the early days of her activities because they feared that she would bring a flood of quackery in her wake. Mrs. Sanger always had a ready answer: "Just as demand and supply are related to all economic questions, so is propaganda a related part of scientific research in the realms of sex psychology. The medical profession will ultimately meet the issue on the demands of public opinion." In a roundabout way which Mrs. Sanger neither intended nor foresaw, that was what happened in the 1930s. Demand for contraceptives outstripped the supply of medical knowledge and gave rise to a huge birth control industry riddled with quackery and dishonesty. On that sorry product of "public opinion," the organized medical profession did finally, though reluctantly, meet the issue.

As early as 1925, the Section on Obstetrics, Gynecology, and Abdominal Surgery of the American Medical Association, at Dickinson's urging, adopted a resolution requesting amendment of state and federal laws "so that physicians may legally give contraceptive information to their patients." The section referred the resolution to the Board of Trustees in 1927, but they handed it back, noting the "great lack of unanimity of opinion" on the subject. Again, in 1932, the association's executive committee refused a proposal to conduct a study of contraception because birth control was "a controversial subject and the committee believes that it would not be advisable at this time to inject this subject before the profession."

But the growing commercial exploitation of contraception continued forcefully "to inject this subject before the profession." At the 1934 session of the association, a Dr. J. D. Brook pointed to the possible dangers in the "innumerable devices for contraception offered to the public." Backed by several supporting resolutions citing the desperate need for study and regulation, Brook demanded a committee to study the medical aspects of contraception. Still the association refused even to look into the matter.

By 1935, the pressure to take some position on birth control had become overwhelming. Mrs. Sanger had propagandized nearly every member of the AMA's House of Delegates. Doctors sympathetic to birth control pleaded for guidance in the ethics of contraceptive practice and especially in choosing among the hundreds of contraceptive products on the market. All doctors demanded action against the flood of questionably effective and often dangerous devices sold to their patients.

Faced with this outcry, the association at last relented and appointed a committee to investigate "contraceptive practices and related problems."

The committee's report in 1936 satisfied no one. While making no judgment on the unethical practices of the contraceptive advertisers, it issued a blast at the relatively respectable birth control organizations. The committee disapproved, it said, "of propaganda directed to the public by lay bodies . . . your committee deplores the support of such agencies by members of the medical profession." With regard to the ethics of indications, the report allowed a long list of medical indications, but dismissed all demographic, eugenic, and especially economic ones. Finally, the report reflected the disappointing state of contraceptive techniques: "Your committee knows of no type of contraception which is reasonably adequate and effective for a large portion of the population . . . No contraceptive technic other than actual continence is intrinsically one hundred per cent safe . . . None are dependable for couples who are intoxicated, subnormal, or lacking in self control." Because of the unsatisfactory character of the report, the association instructed the committee to continue its study for another year.

When the committee reported at the 1937 session of the AMA, it manifested an attitude very different from that of 1936. Perhaps because the Second Circuit Court of Appeals had in the meantime clarified the legal status of contraception, perhaps because Margaret Sanger had announced the disbanding of her propaganda organization—the National Committee for Federal Legislation for Birth Control, perhaps because the AMA committee's membership had undergone some changes, the committee submitted a new report which virtually endorsed birth control. It still insisted that clinics be under strict medical supervision, but it refrained from castigating lay organizations in general. Noting the dangers of increasing commercialization of birth control, the committee also asked that the association "undertake the investigation of materials, devices, and methods recommended or employed for the prevention of conception," and report on these investigations to the profession. In order to overcome widespread medical ignorance of contraception the committee asked the association "to promote thorough instruction in our medical schools with respect to the various factors pertaining to fertility and sterility."

Finally, the committee accepted the position Margaret Sanger had been promoting for two decades: that other than pathologic conditions were valid indications for contraception. Medical opinion on permissible indications had been growing more liberal for some time. Doctors at the clinic raid trial in 1929 testified that physicians had come to realize that birth control should be prescribed to ensure the proper spacing of births. Even in normal women, they pointed out, a close succession of pregnancies could destroy good health. In 1935, the *American Journal of Obstetrics and Gynecology* editorially suggested that "sociologists should establish and announce reasons, if such exist, for giving contraceptive advice to individuals." The statement of the AMA committee went far beyond purely medical, and even, in a sense, sociological, indications. Doctors, the committee advised, should no longer insist that contraception be used only in the treatment or prevention of diagnosable disease; they should honor the good faith of their patients who requested birth control. "Voluntary family limitation," the committee said, "is dependent largely on the judgment and wishes of individual parents."

Though Margaret Sanger regarded the indictment of commercial exploitation and the demand for contraceptive education as important victories, she took greatest pleasure in the committee's liberal position on indications. Contrary views on indications had long divided her from otherwise sympathetic physicians. But she had stood her ground and finally won her point. Her old foe, George Kosmak, conceded in 1940 that "the indications, both medical and social, using the latter term in its wider implications, have been established in a more satisfactory manner." Mrs. Sanger, naturally, had not been the only influence in liberalizing medical opinion on the subject. Broadening definitions of health, largely as a result of public health work, and increased legal latitude played important roles. But Mrs. Sanger virtually alone through the years had insisted that birth control was relevant to those expanding conceptions of health; and she had been instrumental in obtaining the legal decisions that made possible wider medical practice of contraception.

Of course, many physicians had backed birth control at considerable risk to their professional reputations. Robert Latou Dickinson put his prestige on the line when he publicly supported Mrs. Sanger, and he and other doctors often thought that if only she would leave them alone they could muster the support of the profession. To be sure, her controversial presence had frightened off many timid physicians, and her stubbornness had often dashed some of Dickinson's most cherished hopes. But for all Dickinson's own courage and dedication, and for all Mrs. Sanger's often unreasoning intransigence, Dickinson could not escape the fact that his profession might have moved even more slowly had it not been for Margaret Sanger.

When the medical profession at last officially endorsed birth control in 1937, many problems remained, especially the continuing lack of an acceptable technique. But the profession had at last taken up its responsibilities as Margaret Sanger had long desired. Undoubtedly many doctors begrudged her her jubilation. But Dickinson, always the gentleman, gave Mrs. Sanger her due. When she received the Town Hall award in 1937, he wired: "Among foremost health measures originating or developing outside medicine like ether under Morton, microbe hunting under Pasteur, nursing under Nightingale, Margaret Sanger's world wide service holds high rank and is destined eventually to fullest medical recognition."

CHAPTER 7

Medical Nemesis: The Expropriation of Health

Ivan Illich

Ivan Illich is an erudite opponent of industrial society in all its manifestations, and has been described as "the leading Luddite of the twentieth century" (the Luddites were early nineteenth-century English workers who, by destroying manufacturing machinery that threatened their jobs, came to symbolize all attempts to hold back the advance of modern technology). Illich has written books attacking modern education, modern transportation, and modern cities, and in *Medical Nemesis* (1976), he trains his cannon on modern medicine.

Unlike others, who have focused on the increasing problems of high cost and lack of access to high-quality medical care, Illich argues that modern medicine itself is the problem. Our reading here consists of selections taken from two portions in his book. First, selections are presented from the book's overview chapter, "The Epidemics of Modern Medicine." Then comes a piece from a later chapter, "The Medicalization of Life."

Modern doctors, Illich tells us, are no more effective medically than priests. Doctors' reputation for effectiveness has been unearned, he says, because the killer infections that afflicted us at the beginning of the industrial age—tuberculosis, cholera, dysentery, typhoid, and so on— all dwindled before medical intervenion was mounted. Deaths from the major childhood diseases also declined before immunization was widespread. Illich credits nonmedical improvements in the environment, and better nutrition in particular, for the declines in these diseases. Even health measures developed at the instigation of doctors, such as better treatment of water and sewage and wider use of soap, are significant in Illich's view because they moved from the hands of doctors into the hands of the laity, and proved effective when they became a part of a society's culture.

As for the treatment delivered by doctors themselves, Illich claims that it bears no causal relationship to a decline in a given community's rate of disease or increase in average life expectancy. Physicians are found in higher numbers in places in which certain diseases have become rare, but this simply reflects the doctors' preference for places possessing a healthy climate, clean water, and residents with money.

Medical treatment is utterly useless for some infectious diseases and for most noninfectious diseases, Illich maintains. For example, survival rates for the most common types of cancer have remained virtually unchanged over the last twenty-five years. But worse than being merely ineffective, medical treatment itself injures and kills a dismaying number of patients. Illich declares medicine to be a rapidly spreading "epidemic."

Readers will encounter a special set of terms at this point in Illich's discussion. The process by which medicine makes us sick—individually and as a society—is what he terms *iatrogenesis:* The causation of disease by physicians, or by the whole of the medical complex (*iatros* is Greek for "physician," and *genesis* of course means "origin"). He further differentiates three forms: *Clinical* iatrogenesis (the damage done by medical intervention); *social* iatrogenesis (the unnecessary dependence society comes to have on the medical profession and its institutions); and *cultural* iatrogenesis (the sapping of individuals' ability to accept pain, impairment, and death).

The advent of modern medical technology has accompanied the transformation of the physician's role. Doctors have changed from "artisans," who treated patients whom they knew personally, into technicians, who apply scientific rules impersonally. Illich notes that malpractice now has become a technical more than an ethical problem, and human negligence is obscured by blame attributed to machines ("system breakdown") or their absence ("lack of specialized equipment.")

More sophisticated techniques and machines will not solve the problem of iatrogenesis; they will only exacerbate the problem. Illich entitles his book *Medical Nemesis* (Nemesis was the goddess of divine retribution in Greek mythology) because, from his point of view, our modern medical system reflects institutional hubris (extreme arrogance) that brings us only the misfortune of increased ills.

In the final portion selected here, "The Medicalization of Life," the author discusses nonmedical roles physicians play in our culture—as secular priests and magicians—or should play but do not—as ethical leaders who could provide moral leadership for a community by showing compassion for the weak and the troubled.

Readers may well ask, what exactly would Illich wish that physicians do when presented with a sick person who suffers a life-threatening illness? Elsewhere in his book Illich seems to advise that no conventional medical treatment be attempted: Doctors should encourage the sick "to seek a poetic interpretation of their predicament or find an admirable example in some person—long dead or next door—who learned to suffer. Medical procedures multiply disease by moral degradation when they isolate the sick in a professional environment rather than providing society with the motives and disciplines that increase social tolerance for the troubled." The author, we may presume, must be conscious of how provocative such statements seem to others; it would appear to be an open dare for the reader to engage the author in debate.

A few biographical notes about this author may be of interest. Illich has led a truly cosmopolitan life. Born in Austria, he grew up and studied in Europe, then moved to New York City, where he worked for five years as a parish priest in an Irish-Puerto Rican neighborhood, then spent some time in Puerto Rico. He finally settled in Cuernavaca, Mexico.

THE EPIDEMICS OF MODERN MEDICINE

Doctors' Effectiveness—an Illusion

The study of the evolution of disease patterns provides evidence that during the last century doctors have affected epidemics no more profoundly than did priests during earlier times. Epidemics came and went, imprecated by both but touched by neither. They are not modified any more decisively by the rituals performed in medical clinics than by those customary at religious shrines. Discussion of the future of health care might usefully begin with the recognition of this fact.

The infections that prevailed at the outset of the industrial age illustrate how medicine came by its reputation. Tuberculosis, for instance, reached a peak over two generations. In New York in 1812, the death rate was estimated to be higher than 700 per 10,000; by 1882, when Koch first isolated and cultured the bacillus, it had already declined to 370 per 10,000. The rate was down to 180 when the first sanatorium was opened in 1910, even though "consumption" still held second place in the mortality tables. After World War II, but before antibiotics became routine, it had slipped into eleventh place with a rate of 48. Cholera, dysentery, and typhoid similarly peaked and dwindled outside the physician's control. By the time their etiology was understood and their therapy had become specific, these diseases had lost much of their vir-

---

ulence and hence their social importance. The combined death rate from scarlet fever, diphtheria, whooping cough, and measles among children up to fifteen shows that nearly 90 percent of the total decline in mortality between 1860 and 1965 had occurred before the introduction of anti-biotics and widespread immunization. In part this recession may be attributed to improved housing and to a decrease in the virulence of micro-organisms, but by far the most important factor was a higher host-resistance due to better nutrition. In poor countries today, diarrhea and upper-respiratory-tract infections occur more frequently, last longer, and lead to higher mortality where nutrition is poor, no matter how much or how little medical care is available. In England, by the middle of the nineteenth century, infectious epidemics had been replaced by major malnutrition syndromes, such as rickets and pellagra. These in turn peaked and vanished, to be replaced by the diseases of early childhood and, somewhat later, by an increase in duodenal ulcers in young men. When these declined, the modern epidemics took over: coronary heart disease, emphysema, bronchitis, obesity, hypertension, cancer (espe-cially of the lungs), arthritis, diabetes, and so-called mental disorders. Despite intensive research, we have no complete explanation for the genesis of these changes. But two things are certain: the professional practice of physicians cannot be credited with the elimination of old forms of mortality or morbidity, nor should it be blamed for the increased expectancy of life spent in suffering from the new diseases. For more than a century, analysis of disease trends has shown that the environ-ment is the primary determinant of the state of general health of any population. Medical geography, the history of diseases, medical an-thropology, and the social history of attitudes towards illness have shown that food, water, and air, in correlation with the level of socio-political equality and the cultural mechanisms that make it possible to keep the population stable, play the decisive role in determining how healthy grown-ups feel and at what age adults tend to die. As the older causes of disease recede, a new kind of malnutrition is becoming the most rapidly expanding modern epidemic. One-third of humanity sur-vives on a level of undernourishment which would formerly have been lethal, while more and more rich people absorb ever greater amounts of poisons and mutagens in their food. Some modern techniques, often developed with the help of doctors, and optimally effective when they become part of the culture and environment or when they are applied independently of professional delivery, have also effected changes in general health, but to a lesser degree. Among these can be included contraception, smallpox vaccination of infants, and such nonmedical health measures as the treatment of water and sewage, the use of soap and scissors by midwives, and some antibacterial and insecticidal procedures. The importance of many of these practices was first recognized and stated by doctors—often courageous dissidents who suffered for their recommendations—but this does not consign soap, pincers, vaccination needles, delousing preparations, or condoms to the category of "medical equipment." The most recent shifts in mortality from younger to older groups can be explained by the incorporation of these procedures and devices into the layman's culture. In contrast to environmental improvements and modern nonprofessional health measures, the specifically medical treatment of people is never significantly related to a decline in the compound disease burden or to a rise in life expectancy. Neither the proportion of doctors in a population nor the clinical tools at their disposal nor the number of hospital beds is a causal factor in the striking changes in over-all patterns of disease. The new techniques for recognizing and treating such conditions as pernicious anemia and hypertension, or for correcting congenital malformations by surgical intervention, redefine but do not reduce morbidity. The fact that the doctor population is higher where certain diseases have become rare has little to do with the doctors' ability to control or eliminate them. It simply means that doctors deploy themselves as they like, more so than other professionals, and that they tend to gather where the climate is healthy, where the water is clean, and where people are employed and can pay for their services.

Useless Medical Treatment

Awe-inspiring medical technology has combined with egalitarian rhetoric to create the impression that contemporary medicine is highly effective. Undoubtedly, during the last generation, a limited number of specific procedures have become extremely useful. But where they are not monopolized by professionals as tools of their trade, those which are applicable to widespread diseases are usually very inexpensive and require a minimum of personal skills, materials, and custodial services from hospitals. In contrast, most of today's skyrocketing medical expenditures are destined for the kind of diagnosis and treatment whose effectiveness at best is doubtful. To make this point I will distinguish between infectious and noninfectious diseases.

In the case of infectious diseases, chemotherapy has played a significant role in the control of pneumonia, gonorrhea, and syphilis. Death from pneumonia, once the "old man's friend," declined yearly by 5 to 8 percent after sulphonamides and antibiotics came on the market. Syphilis, yaws, and many cases of malaria and typhoid can be cured quickly and easily. The rising rate of venereal disease is due to new mores, not to ineffectual medicine. The reappearance of malaria is due to the development of pesticide-resistant mosquitoes and not to any lack of new antimalarial drugs. Immunization has almost wiped out paralytic poliomyelitis, a disease of developed countries, and vaccines have certainly contributed to the decline of whooping cough and measles, thus seeming to confirm the popular belief in "medical progress." But for most other infections, medicine can show no comparable results. Drug treatment has helped to reduce mortality from tuberculosis, tetanus, diphtheria, and scarlet fever, but in the total decline of mortality or morbidity from these diseases, chemotherapy played a minor and possibly insignificant role. Malaria, leishmaniasis, and sleeping sickness indeed receded for a time under the onslaught of chemical attack, but are now on the rise again.

The effectiveness of medical intervention in combatting noninfectious diseases is even more questionable. In some situations and for some conditions, effective progress has indeed been demonstrated: the partial prevention of caries through fluoridation of water is possible, though at a cost not fully understood. Replacement therapy lessens the direct impact of diabetes, though only in the short run. Through intravenous feeding, blood transfusions, and surgical techniques, more of those who get to the hospital survive trauma, but survival rates for the most common types of cancer—those which make up 90 percent of the cases—have remained virtually unchanged over the last twenty-five years. This fact has consistently been clouded by announcements from the American Cancer Society reminiscent of General Westmoreland's proclamations from Vietnam. On the other hand, the diagnostic value of the Papanicolaou vaginal smear test has been proved: if the tests are given four times a year, early intervention for cervical cancer demonstrably increases the five-year survival rate. Some skin-cancer treatment is highly effective. But there is little evidence of effective treatment of most other cancers. The five-year survival rate in breast-cancer cases is 50 percent, regardless of the frequency of medical check-ups and regardless of the treatment used. Nor is there evidence that the rate differs from that among untreated women. Although practicing doctors and the publicists of the medical establishment stress the importance of early detection and treatment of this and several other types of cancer, epidemiologists have begun to doubt that early intervention can alter the rate of survival. Surgery and chemotherapy for rare congenital and rheumatic heart disease have increased the chances for an active life for some of those who suffer from degenerative conditions. The medical treatment of common cardiovascular disease and the intensive treatment of heart disease, however, are effective only when rather exceptional circumstances combine that are outside the physician's control. The drug treatment of high blood pressure is effective and warrants the risk of side-effects in the few in whom it is a malignant condition; it represents a considerable risk of serious harm, far outweighing any proven benefit, for the 10 to 20 million Americans on whom rash artery-plumbers are trying to foist it.

Doctor-Inflicted Injuries

Unfortunately, futile but otherwise harmless medical care is the least important of the damages a proliferating medical enterprise inflicts on contemporary society. The pain, dysfunction, disability, and anguish resulting from technical medical intervention now rival the morbidity due to traffic and industrial accidents and even war-related activities, and make the impact of medicine one of the most rapidly spreading epidemics of our time. Among murderous institutional torts, only modern malnutrition injures more people than iatrogenic disease in its various manifestations. In the most narrow sense, iatrogenic disease includes only illnesses that would not have come about if sound and professionally recommended treatment had *not* been applied. Within this definition, a patient could sue his therapist if the latter, in the course of his management, failed to apply a recommended treatment that, in the physician's opinion, would have risked making him sick. In a more general and more widely accepted sense, clinical iatrogenic disease comprises all clinical conditions for which remedies, physicians, or hospitals are the pathogens, or "sickening" agents. I will call this plethora of therapeutic side-effects *clinical iatrogenesis*. They are as old as medicine itself, and have always been a subject of medical studies.

Medicines have always been potentially poisonous, but their unwanted side-effects have increased with their power and widespread use. Every twenty-four to thirty-six hours, from 50 to 80 percent of adults in the United States and the United Kingdom swallow a medically prescribed chemical. Some take the wrong drug; others get an old or a contaminated batch, and others a counterfeit; others take several drugs in dangerous combinations, and still others receive injections with improperly sterilized syringes. Some drugs are addictive, others mutilating, and others mutagenic, although perhaps only in combination with food coloring or insecticides. In some patients, antibiotics alter the normal bacterial flora and induce a superinfection, permitting more resistant organisms to proliferate and invade the host. Other drugs contribute to the breeding of drug-resistant strains of bacteria. Subtle kinds of poisoning thus have spread even faster than the bewildering variety and ubiquity of nostrums. Unnecessary surgery is a standard procedure. *Disabling nondiseases* result from the medical treatment of nonexistent diseases and are on the increase: the number of children disabled in

Massachusetts through the treatment of cardiac nondisease exceeds the number of children under effective treatment for real cardiac disease.

Doctor-inflicted pain and infirmity have always been a part of medical practice. Professional callousness, negligence, and sheer incompetence are age-old forms of malpractice. With the transformation of the doctor from an artisan exercising a skill on personally known individuals into a technician applying scientific rules to classes of patients, malpractice acquired an anonymous, almost respectable status. What had formerly been considered an abuse of confidence and a moral fault can now be rationalized into the occasional breakdown of equipment and operators. In a complex technological hospital, negligence becomes "random human error" or "system breakdown," callousness becomes "scientific detachment," and incompetence becomes "a lack of specialized equipment." The depersonalization of diagnosis and therapy has changed malpractice from an ethical into a technical problem.

In 1971, between 12,000 and 15,000 malpractice suits were lodged in United States courts. Less than half of all malpractice claims were settled in less than eighteen months, and more than 10 percent of such claims remain unsettled for over six years. Between sixteen and twenty percent of every dollar paid in malpractice insurance went to compensate the victim; the rest was paid to lawyers and medical experts. In such cases, doctors are vulnerable only to the charge of having acted against the medical code, of the incompetent performance of prescribed treatment, or of dereliction out of greed or laziness. The problem, however, is that most of the damage inflicted by the modern doctor does not fall into any of these categories. It occurs in the ordinary practice of well-trained men and women who have learned to bow to prevailing professional judgment and procedure, even though they know (or could and should know) what damage they do.

The United States Department of Health, Education, and Welfare calculates that 7 percent of all patients suffer compensable injuries while hospitalized, though few of them do anything about it. Moreover, the frequency of reported accidents in hospitals is higher than in all industries but mines and high-rise construction. Accidents are the major cause of death in American children. In proportion to the time spent there, these accidents seem to occur more often in hospitals than in any other kind of place. One in fifty children admitted to a hospital suffers an accident which requires specific treatment. University hospitals are relatively more pathogenic, or, in blunt language, more sickening. It has also been established that one out of every five patients admitted to a typical research hospital acquires an iatrogenic disease, sometimes trivial, usually requiring special treatment, and in one case in thirty leading to death. Half of these episodes result from complications of drug therapy; amazingly, one in ten comes from diagnostic procedures. Despite good intentions and claims to public service, a military officer with a similar record of performance would be relieved of his command, and a restaurant or amusement center would be closed by the police. No wonder that the health industry tries to shift the blame for the damage caused onto the victim, and that the dope-sheet of a multinational pharmaceutical concern tells its readers that "iatrogenic disease is almost always of neurotic origin."

Defenseless Patients

The undesirable side-effects of approved, mistaken, callous, or contraindicated technical contacts with the medical system represent just the first level of pathogenic medicine. Such *clinical iatrogenesis* includes not only the damage that doctors inflict with the intent of curing or of exploiting the patient, but also those other torts that result from the doctor's attempt to protect himself against the possibility of a suit for malpractice. Such attempts to avoid litigation and prosecution may now do more damage than any other iatrogenic stimulus.

On a second level, medical practice sponsors sickness by reinforcing a morbid society that encourages people to become consumers of curative, preventive, industrial, and environmental medicine. On the one hand defectives survive in increasing numbers and are fit only for life under institutional care, while on the other hand, medically certified symptoms exempt people from industrial work and thereby remove them from the scene of political struggle to reshape the society that has made them sick. Second-level iatrogenesis finds its expression in various symptoms of social overmedicalization that amount to what I shall call the expropriation of health. This second-level impact of medicine I designate as *social iatrogenesis,* and I shall discuss it in Part II.

On a third level, the so-called health professions have an even deeper, culturally health-denying effect insofar as they destroy the potential of people to deal with their human weakness, vulnerability, and uniqueness in a personal and autonomous way. The patient in the grip of contemporary medicine is but one instance of mankind in the grip of its pernicious techniques. This *cultural iatrogenesis,* which I shall discuss in Part III, is the ultimate backlash of hygienic progress and consists in the paralysis of healthy responses to suffering, impairment, and death. It occurs when people accept health management designed on the engineering model, when they conspire in an attempt to produce, as if it were a commodity, something called "better health." This inevitably results in the managed maintenance of life on high levels of sublethal illness. This ultimate evil of medical "progress" must be clearly distinguished from both clinical and social iatrogenesis.

I hope to show that on each of its three levels iatrogenesis has become medically irreversible: a feature built right into the medical endeavor. The unwanted physiological, social, and psychological by-products of diagnostic and therapeutic progress have become resistant to medical remedies. New devices, approaches, and organizational arrangements, which are conceived as remedies for clinical and social iatrogenesis, themselves tend to become pathogens contributing to the new epidemic. Technical and managerial measures taken on any level to avoid damaging the patient by his treatment tend to engender a self-reinforcing iatrogenic loop analogous to the escalating destruction generated by the polluting procedures used as antipollution devices.

I will designate this self-reinforcing loop of negative institutional feedback by its classical Greek equivalent and call it *medical nemesis*. The Greeks saw gods in the forces of nature. For them, nemesis represented divine vengeance visited upon mortals who infringe on those prerogatives the gods enviously guard for themselves. Nemesis was the inevitable punishment for attempts to be a hero rather than a human being. Like most abstract Greek nouns, Nemesis took the shape of a divinity. She represented nature's response to *hubris:* to the individual's presumption in seeking to acquire the attributes of a god. Our contemporary hygienic hubris has led to the new syndrome of medical nemesis.

By using the Greek term I want to emphasize that the corresponding phenomenon does not fit within the explanatory paradigm now offered by bureaucrats, therapists, and ideologues for the snowballing diseconomies and disutilities that, lacking all intuition, they have engineered and that they tend to call the "counterintuitive behavior of large systems." By invoking myths and ancestral gods I should make it clear that my framework for analysis of the current breakdown of medicine is foreign to the industrially determined logic and ethos. I believe that the *reversal of nemesis* can come only from within man and not from yet another managed (heteronomous) source depending once again on presumptuous expertise and subsequent mystification.

Medical nemesis is resistant to medical remedies. It can be reversed only through a recovery of the will to self-care among the laity, and through the legal, political, and institutional recognition of the right to care, which imposes limits upon the professional monopoly of physicians. My final chapter proposes guidelines for stemming medical nemesis and provides criteria by which the medical enterprise can be kept within healthy bounds. I do not suggest any specific forms of health care or sick-care, and I do not advocate any new medical philosophy any more than I recommend remedies for medical technique, doctrine, or organization. However, I do propose an alternative approach to the use of medical organization and technology together with the allied bureaucracies and illusions.

THE MEDICALIZATION OF LIFE

Black Magic

Technical intervention in the physical and biochemical make-up of the patient or of his environment is not, and never has been, the sole function of medical institutions. The removal of pathogens and the application of remedies (effective or not) are by no means the sole way of mediating between man and his disease. Even in those circumstances in which the physician is technically equipped to play the technical role to which he aspires, he inevitably also fulfills religious, magical, ethical, and political functions. In each of these functions the contemporary physician is more pathogen than healer or just anodyne.

Magic or healing through ceremonies is clearly one of the important traditional functions of medicine. In magic the healer manipulates the setting and the stage. In a somewhat impersonal way he establishes an ad hoc relationship between himself and a group of individuals. Magic works if and when the intent of patient and magician coincides, though it took scientific medicine considerable time to recognize its own practitioners as part-time magicians. To distinguish the doctor's professional exercise of white magic from his function as engineer (and to spare him the charge of being a quack), the term "placebo" was created. Whenever a sugar pill works because it is given by the doctor, the sugar pill acts as a placebo. A placebo (Latin for "I will please") pleases not only the patient but the administering physician as well.

In high cultures, religious medicine is something quite distinct from magic. The major religions reinforce resignation to misfortune and offer a rationale, a style, and a community setting in which suffering can become a dignified performance. The opportunities offered by the acceptance of suffering can be differently explained in each of the great traditions: as karma accumulated through past incarnations; as an invitation to Islam, the surrender to God; or as an opportunity for closer association with the Savior on the Cross. High religion stimulates personal responsibility for healing, sends ministers for sometimes pompous and sometimes effective consolation, provides saints as models, and usually provides a framework for the practice of folk medicine. In our kind of secular society religious organizations are left with only a small part of their former ritual healing roles. One devout Catholic might derive intimate strength from personal prayer, some marginal groups of recent arrivals in São Paolo might routinely heal their ulcers in Afro-Latin dance cults, and Indians in the valley of the Ganges still seek health in the singing of the Vedas. But such things have only a remote parallel in societies beyond a certain per capita GNP. In these industrialized societies secular institutions run the major myth-making ceremonies.

The separate cults of education, transportation, and mass communication promote, under different names, the same social myth which Voeglin describes as contemporary gnosis. Common to a gnostic worldview and its cult are six characteristics: (1) it is practiced by members of a movement who are dissatisfied with the world as it is because they see it as intrinsically poorly organized. Its adherents are (2) convinced that salvation from this world is possible (3) at least for the elect and (4) can be brought about within the present generation. Gnostics further believe that this salvation depends (5) on technical actions which are reserved (6) to initiates who monopolize the special formula for it. All these religious beliefs underlie the social organization of technological medicine, which in turn ritualizes and celebrates the nineteenth-century ideal of progress.

Among the important nontechnical functions of medicine, a third one is ethical rather than magical, secular rather than religious. It does not depend on a conspiracy into which the sorcerer enters with his adept, nor on myths to which the priest gives form, but on the shape which medical culture gives to interpersonal relations. Medicine can be so organized that it motivates the community to deal in a more or less personal fashion with the frail, the decrepit, the tender, the crippled, the depressed, and the manic. By fostering a certain type of social character, a society's medicine could effectively lessen the suffering of the diseased by assigning an active role to all members of the community in the compassionate tolerance for and the selfless assistance of the weak. Medicine could regulate society's gift relationships. Cultures where compassion for the unfortunate, hospitality for the crippled, leeway for the troubled, and respect for the old have been developed can, to a large extent, integrate the majority of their members into everyday life.

Healers can be priests of the gods, lawgivers, magicians, mediums, barber-pharmacists, or scientific advisers. No common name with even the approximate semantic range covered by our "doctor" existed in Europe before the fourteenth century. In Greece the repairman, used mostly for slaves, was respected early, though he was not on a level with the healing philosopher or even with the gymnast for the free. Republican Rome considered the specialized curers a disreputable lot. Laws on water supply, drainage, garbage removal, and military training, combined with the state cult of healing gods, were considered sufficient; grandmother's brew and the army sanitarian were not dignified by special attention. Until Julius Caesar gave citizenship to the first group of Asclepiads in 46 B.C., this privilege was refused to Greek physicians and healing priests. The Arabs honored the physician, the Jews left health care to the quality of the ghetto or, with a bad conscience, brought in the Arab physician. Medicine's several functions combined in different ways in different roles. The first occupation to monopolize health care is that of the physician of the late twentieth century.

Paradoxically, the more attention is focused on the technical mastery of disease, the larger becomes the symbolic and nontechnical function performed by medical technology. The less proof there is that more money increases survival rates in a given branch of cancer treatment, the more money will go to the medical divisions deployed in that special theater of operations. Only goals unrelated to treatment, such as jobs for the specialists, equal access by the poor, symbolic consolation for patients, or experimentation on humans, can explain the expansion of lung-cancer surgeries during the last twenty-five years. Not only white coats, masks, antiseptics, and ambulance sirens but entire branches of medicine continue to be financed because they have been invested with nontechnical, usually symbolic power.

Willy-nilly the modern doctor is thus forced into symbolic, nontechnical roles. Nontechnical functions prevail in the removal of adenoids: more than 90 percent of all tonsillectomies performed in the United States are technically unnecessary, yet 20 to 30 percent of all children still undergo the operation. One in a thousand dies directly as a consequence of the operation and 16 in a thousand suffer from serious complications. All lose valuable immunity mechanisms. All are subjected to emotional aggression: they are incarcerated in a hospital, separated from their parents, and introduced to the unjustified and more often than not pompous cruelty of the medical establishment. The child learns to be exposed to technicians who, in his presence, use a foreign language in which they make judgments about his body; he learns that his body may be invaded by strangers for reasons they alone know; and he is made to feel proud to live in a country where social security pays for such a medical initiation into the reality of life. . . .

CHAPTER 8

The Car Culture

James J. Flink

We use many products of technology daily without much thought or emotion. The convenience of hot and cold running water, for example, is wonderful to have, but it does not hold for the user much in the way of symbolic or emotional significance. The plumbing does its job, and that's that.

This is less so in the case of automobiles, with which Americans are said to have had a long-standing romance. *The Car Culture* covers the period 1900 to 1930, when automobiles became a seemingly indispensable part of American life and were viewed as much more than simply a means of transportation.

In surveying the far-reaching impact of "automobility" on the country, James Flink notes economic, social, moral, and environmental changes set in motion by the automobile. The economic impact extended beyond the automobile manufacturers and their dealers—other industries, such as steel and petroleum, underwent revolutionary change in response to the dramatic growth of the auto industry, and government at all levels also had to respond with road construction and taxes to pay for them.

The economic impact on a family's budget was also of course considerable, and the author tells us of the extreme lengths to which working-class families went in order to purchase cars. Advertising fanned demand, and installment financing came to be accepted—despite the uneasiness of many bankers and some industrialists, including Henry Ford, who were alarmed about consumers' abandonment of the habits of thrift.

Collecting taxes for expanded government services did not prove to be a problem—the almost cheerful willingness of motorists to pay higher registration fees and gasoline taxes testifies to the soaring enthusiasm car owners had for automobility. The favorable reception to cars in this period is found in the South as well as in the North, and extended across lines of race and class.

Flink relies upon the observations of Robert and Helen Lynd, the husband-wife sociologists whose extended studies of the pseudonymous Middletown (actually, Muncie, Indiana) provide a closely documented look at all aspects of life in a midwestern city during the 1920s and 1930s. One of the striking findings of the Lynds was that the automobile's dominance of city life did not seem to be slowed by the onset of the Depression.

It was the Lynds who also noticed that the automobile was symbolically more important to the working class than to the middle class. Car ownership was viewed as a key element in realizing the "American dream," and this conviction remained right through the hard years of the 1930s.

One of the central questions posed by the automobile invasion concerns its impact on community, on the individual's relationship to his or her neighborhood or town. Flink credits the automobile with challenging the repressive side of a given community; the mobility conferred by the automobile allowed individuals to escape much more easily the pressure of group conformity or prejudice and start life anew elsewhere. But what Flink views as the most positive aspects of community were also undermined by the automobile. Workers came to live in scattered neighborhoods away from their place of employment and their co-workers. Family and neighbors in Middletown no longer spent their summer evenings on the porch visiting and socializing.

Early automobile advertising promoted the idea that car ownership strengthened family togetherness, but Flink argues emphatically that in fact automobiles did not bring families together. He stops short of attributing the rising divorce rates of the 1920s to the automobile, but he does remind us that the claim of early enthusiasts, that the family car would slow the divorce rate, was not borne out. Teenagers and their parents did not seem to be brought closer together than husbands and wives. The car provided teens with new mobility, which undercut parental authority, and, at least at the time, there was much commentary about the new opportunities for illicit sex the automobile provided. Flink registers a skeptical note about whether this worry about sex had any objective foundation.

In the end, Flink remains unconvinced that the automobile brought great advantages to American society. He points to the results of a 1920 survey of car owners, in which few respondents claimed that car ownership had actually improved the quality of their lives. He notes the undermining of community and family and the accompanying increase in anonymity and anomie (feeling lost, lonely, without direction). He stresses early examples of prescient individuals who anticipated the traffic and parking problems that automobiles would create.

With hindsight, Flink says, it can be seen that American politicians and urban planners made a terrible mistake when they chose for cities the course of unlimited accommodation to the automobile, instead of encouraging bus, rail, and other modes of transportation. One later consequence of their decision was the subdivision of urban America into urban ghettos and middle-class suburbs.

This problem was not anticipated in the 1920s, nor was the air pollution that would be created in cities choked with auto exhaust. But even in this early period in automotive history, there was considerable concern about the terrible toll of injuries and deaths automobiles brought. Flink is unsparing of the automobile industry, which sold cars that were overpowered and which in design gave styling considerations precedence over safety.

Today, we are trying to undo some of the damage. Federal automobile emissions standards have pushed manufacturers to redesign engines. Federal funding of urban mass transportation boomed in the 1970s (but dropped to a trickle in the 1980s). We remain, nevertheless, a country dominated by private automobiles whose collective exhaust palpably fouls our air. We continue to live a distance, in many cases a considerable distance, from the places in which we work. We have learned that it is much easier to purchase a fleet of new buses than it is to attract commuters to ride them. And we have seen that the romance with the automobile brooks no interference: A majority of Americans appeared to view the 55 mile-per-hour speed limit as an infringement of their sovereign right to enjoy speed. Ours remains very much a car culture.

THE AUTOMOBILE CULTURE: PROSPECTS AND PROBLEMS

With his characteristic flair for understating the spectacular, President Warren G. Harding said in his April 12, 1921, message to Congress that "the motorcar has become an indispensable instrument in our political, social, and industrial life." An anonymous resident of Muncie, Indiana, was more accurate when he exclaimed to the Lynds, "Why on earth do you need to study what's changing this country? I can tell you what's happening in just four letters: A-U-T-O!" A Middle-

SOURCE: From *The Car Culture* by James J. Flink. Copyright © 1975 by Massachusetts Institute of Technology. Reprinted by permission of MIT Press.

town Bible class teacher was told by her class that the only animal God created that man could get along without was the horse. "Ten or twelve years ago a new horse fountain was installed at the Courthouse square; now it remains dry during most of the blazing heat of a Mid-Western summer, and no one cares," wrote the Lynds in 1929. "The 'horse culture' of Middletown has almost disappeared. Nor was the horse culture in all the years of its undisputed sway ever as pervasive a part of the life of Middletown as is the cluster of habits that have grown up overnight around the automobile."

During the 1920s automobility became the backbone of a new consumer-goods-oriented society and economy that has persisted into the present. By the mid-1920s automobile manufacturing ranked first in value of product and third in value of exports among American industries. In 1926 United States motor vehicle factory sales had a wholesale value of over $3 billion and American motorists spent over $10 billion that year in operating expenses to travel some 141 billion miles. The automobile industry was the lifeblood of the petroleum industry, one of the chief customers of the steel industry, and the biggest consumer of many other industrial products, including plate glass, rubber, and lacquers. The technologies of these ancillary industries, particularly steel and petroleum, were revolutionized by the new demands of motorcar manufacturing. The construction of streets and highways was the second largest item of governmental expenditure during the 1920s. The motorcar was responsible for a suburban real estate boom and for the rise of many new small businesses, such as service stations and tourist accommodations. In 1929, the last year of the automobility induced boom, the 26.7 million motor vehicles registered in the United States (one for every 4.5 persons) traveled an estimated 198 billion miles, and in that year alone government spent $2.237 billion on roads and collected $849 million in special motor vehicle taxes. The eminent social and economic historian Thomas C. Cochran noted the central role of automobility and concluded: "No one has or perhaps can reliably estimate the vast size of capital invested in reshaping society to fit the automobile. Such a figure would have to include expenditures for consolidated schools, suburban and country homes, and changes in business location as well as the more direct investment mentioned above. This total capital investment was probably the major factor in the boom of the 1920s, and hence in the glorification of American business."

The automobile culture became truly national in the 1920s as significant early regional differences in automobility lessened. California led the nation in 1929 as it had in 1910 in ratio of population to motor vehicle registrations. It remained true as well that the leading regions in motor vehicles per capita were still the Pacific and West North Central states and that the South continued to lag behind the rest of the country in adopting the automobile. But the gap among the various regions of the United States had closed appreciably by 1920. During the decade 1910-1920 automobile registrations per capita increased more rapidly in the Mountain states and in the South, the early laggards in adopting the automobile, than in the East North Central, Middle Atlantic, and New England states. Although the agricultural states of the trans-Mississippi West continued to be the best market for new cars, and California remained known in the industry as "a bottomless pit" for automobile sales, there was evidence that regional differences in the diffusion of the motorcar were becoming less significant. With a United States average of 10.1 persons for every motor vehicle in 1921, California ranked first with a ratio of 5.2:1 and Mississippi last with 27.5:1. By 1929 the United States average was 4.5:1. California still led the states with 2.3:1, and Alabama ranked last with 9:1. Probably the range of variation would be less among the states if one adequately adjusted for such crucial demographic variables as family size, age, and sex distribution.

Another indication that the automobile culture had become national by 1920 was that automobile ownership was beginning to approximate the distribution of population along a rural-urban dimension. By 1924 the 53.1 percent of the population that lived on farms and in cities and towns under 5,000 population owned 50.3 percent of the motor vehicles; the 21 percent in cities between 5,000 and 100,000 population owned 28.2 percent; and the 25.9 percent in cities over 100,000 population owned 21.5 percent. A survey of car ownership among over 4.1 million American families conducted in 1927 by the General Federation of Women's Clubs showed only slight differences in the percentage of families owning automobiles in cities of various sizes. The range was from a low of 54 percent of families owning cars in cities of 100,000 population and over, to a progressively increasing percentage as cities got smaller, which peaked at 60.5 percent for towns under 1,000 population. The survey showed that 55.7 percent of the 27.5 million families in the United States in 1927 owned automobiles and that 2.7 million of these families (18 percent of those owning automobiles) owned two or more cars.

The only thing that prevented the American automobile culture of the 1920s from being universally shared was the inequitable income distribution of Coolidge prosperity. It seems significant, although to my knowledge it has never before been stressed, that almost half of American families still did not own a car in 1927, the year when the market for new cars became saturated and the replacement demand for cars for the first time exceeded the demand from initial purchasers and multiple-car owners combined. *Motor* calculated in 1921 that ownership of a $600 automobile necessitated an annual income of at least $2,800 if one lived in a city and at least $1,936 if one lived in the country. In 1924 the National Automobile Chamber of Commerce (NACC) estimated that "the entire field of those receiving under $1,500 [in income] yearly is still unsupplied with motor transportation." The NACC thought that "the growth in the motor vehicle market depends on the ability of the lower income brackets to purchase *used cars*, not necessarily new ones" and that industry policy ought to be "pouring [new cars] in at the top, with the used cars being traded in and going to a secondary market."

The market for automobiles among the lower-income brackets by 1924 was thus conceived by the industry's main trade association primarily as a dumping ground to solve what was known as "the used car problem." From the beginnings of the industry, most automobile manufacturers and dealers had encouraged owners to trade up to higher-priced models or to keep up with technological improvements and changing fashions by replacing their cars long before they were ready for the junk heap. Owners in turn expected a generous trade-in allowance. The problem was that the depreciation was great because used car prices were fixed at the maximal amounts that new, less affluent classes of purchasers could afford to pay. As early as 1906 *Motor World* observed that "the car that sold for $2,000 when new will seldom bring much more than half that price when a year old and at the end of its second year this will practically be halved again despite the fact that as a well-built piece of machinery it may have several years of efficient life before it." The automobile dealer was therefore increasingly faced with having to sell several used cars at a loss in order to make one new car sale. By the mid-1920s, when the market for cars was becoming saturated, most dealers were selling more used cars than new models. Used car prices had been standardized as early as 1914 by the "Used Car Central Market Report," published by the Chicago Automobile Trade Association, which was the forerunner of the now familiar dealers' "Blue Book." By the early 1920s the public had been educated "that they must charge off depreciation to transportation and not expect too much for their old cars." But the trade-in remained a losing proposition until the Ford Motor Company in 1925 and General Motors in 1928 inaugurated plans to help their dealers dispose of used cars at a slight profit.

Although even used cars that brought a mere $10 to $25 profit to the automobile dealer were beyond the reach of almost half of all American families in 1927, still other households operating on marginal incomes skirted insolvency and sacrificed essentials for personal automobility. *Motor* noted that only 357,598 Americans in 1914 earned enough money to pay federal income taxes, but 1,287,784 Americans owned motorcars. As early as 1907 *Horseless Age* had expressed concern that "extravagance is reckless and something must be done before utter ruin follows in the wake of folly. . . . Many owners of houses worth from $5,000 to $15,000, which they have acquired after years of toil, are mortgaging them in order to buy automobiles." The Lynds found in 1929 that many working-class families in Middletown were mortgaging homes to buy cars, that "a working man earning $35.00 a week frequently plans to use one week's pay each month as payment for his car," and that "the automobile has apparently unsettled the habit of careful saving for some families." "We'd rather do without clothes than give up the car," a working-class mother of nine told the Lynds. "I'll go without food before I'll see us give up the car," said another working-class wife emphatically. "Meanwhile," observed the Lynds, "advertisements pound away at Middletown people with the tempting advice to spend money for automobiles for the sake of their homes and families." One such automobile advertisement urged Middletowners: "Hit the trail to better times!"

Advertising undoubtedly played an important role in booming automobile sales beyond the bounds of sanity in the 1920s. The automobile industry during the decade became "one of the heaviest users of magazine space as well as of newspaper space and of other types of mediums. . . . the expansion in the advertising of passenger cars and accessories, parts and supplies in 1923 being particularly noteworthy." Expenditures for automobile advertising in magazines alone climbed from $3.5 million in 1921 to $6.2 million in 1923 and to $9.269 million in 1927.

Automobile advertisements incredibly managed to make irresistible to an increasing number of American families in the 1920s a combination of time payments on the family car, a mortgaged home on the outskirts of town, and the inevitable traffic jam that accompanied the Sunday afternoon drive. As the utilitarian virtues of the product came to be taken for granted by the consumer, automobile advertisements gave more emphasis to making the consumer style conscious and to psychological inducements to buy cars. Especially potent and pervasive themes remained the fusing of rural and urban advantages and family togetherness. The first played on deep internalization of the agrarian myth among Americans. As *Automobile* explained, "Holding the family together is with many families one of the strongest arguments that can be advanced in these days when the divorce mill is grinding overtime. . . . It is an argument that has sold thousands of cars and will continue to sell tens of thousands more, even if salesmen never mention it. Any force that makes stronger the family ties is bound to be a potent force; it is bound to be an enduring force. With the middle classes it is a very strong force."

Paying for mass personal automobility presented problems. By 1910, conservative country bankers were beginning to complain about the growing tendency of farmers to withdraw their savings to purchase motorcars. "Who made the banks dictators?" responded *Motor Age*. "If bank presidents and cashiers are to criticize every so-called luxury or pleasure that to some extent depletes their deposits, it would perhaps be best for Congress to make them guardians of the public funds and permit them, as some do at present, to invest their deposits in most questionable enterprises." *Motor World* was also incensed in 1910 that the big bond houses were refusing to bid on municipal bonds of cities in the Middle West that had "too many automobiles in proportion to population." The New York bond house of Spencer Trask & Company explained in a bulletin, however, that "our people, never of a particularly economical disposition, have been carried away by the automobile craze, and thousands are running cars who cannot afford to do so without mortgaging property, while thousands of others are now investing in motors who formerly invested in bonds. It is calculated that upward of $300,000,000 will be absorbed by the automobile industry this year [1910], which represents the interest on about two-thirds of our entire prospective crops of the present year. This is a phase in our political economy which deserves more consideration than is usually given it."

The early so-called "mortgage on the farm to buy a car scare" was most effectively countered by Benjamin Briscoe. He sent out a questionnaire to some 24,000 bankers across the country and received 4,830 replies. The results showed that of the total 198,216 motorcars in the bankers' towns, mortgages had been placed to buy only 1,254 cars and money borrowed without mortgage to buy another 7,475. What the trade journals' analyses of these data underplayed, however, was that the market for new cars in 1910 was still far from being saturated among people who could well afford personal automobility. *Horseless Age* estimated, for example, that although only about 400,000 automobiles had been produced up to that time, almost a million American families had incomes above $3,000 a year. With the coming of the mass-produced car in the next decade, this market would become saturated, the industry would begin to encourage less affluent classes of purchasers to become automobile owners through the extension of consumer installment credit, and the worst fears of the conservative bankers would come to be realized.

The Morris Plan banks began financing time sales of automobiles in 1910. In 1911 the Studebaker Corporation announced that it would accept notes endorsed by its dealers on the purchases of EMF and Flanders cars. Walter E. Flanders, the Studebaker general manager, explained, "We have in view the future rather than the immediate present. . . . We have considered the advent of the credit in this business as inevitable and our move is but the consummation of a plan long since laid." By 1912, commercial vehicles were commonly sold on time, and a few large automobile dealers had inaugurated their own plans for the installment selling of passenger cars.

The manufacturers of moderately priced cars, under pressure from their dealers, by 1915 came to see installment selling as an alternative to Henry Ford's strategy of progressively lowering the price of the Model T. On November 8, 1915, the Guaranty Securities Company was formed in Toledo, Ohio, with John N. Willys and Alfred P. Sloan, Jr., among its directors, to finance time sales of Overland and Willys-Knight cars. Reorganized as the Guaranty Securities Corporation of New York City, by April 1, 1916, the company was financing time sales of twenty-one makes of car, including all General Motors lines, Dodge, Ford, Hudson, Maxwell, REO, and Studebaker. General Motors, as noted, was the first automobile manufacturer directly to finance time sales of its products with the creation of the General Motors Acceptance Corporation on March 15, 1919. By the spring of 1921, over 110 automobile finance corporations were in existence, and about 50 percent of all automobiles were being sold on some form of deferred payment system.

Bankers continued to oppose the installment selling of automobiles. At a meeting of the Chicago Central Automobile Credit Association on March 20, 1922, for example, the principal speaker focused on the "widespread conviction on the part of the bankers that the financing of automobile purchases was a more or less hazardous business and that furthermore the operation of the financing companies had a tendency to encourage extravagance and was contrary to the thrift ideas which the bankers were so anxious to establish in the minds of the public." Ironically, Henry Ford, who generally had little respect for bankers, agreed with them on this point. He advised the Wisconsin Bankers' Association at its 1915 annual convention to adopt the slogan, "Get Cash, Pay Cash." It was brought out that 90 percent of a $70 million dollar investment in motorcars in Wisconsin represented money withdrawn from savings, money borrowed, or notes purchased. "It has always seemed to me that this putting off the day of payment for anything but permanent improvements was a fundamental mistake," said Ford. "The Ford Motor Company is not interested in promulgating any plan which extends credits for motorcars or for anything else." Ransom E. Olds felt the same way. Upon his resignation as president in 1923 to become chairman of the REO Motor Car Company board of directors, Olds assailed the installment selling plan adopted by other automobile manufacturers. "I believe that the plan is dangerous and a menace to the business," said Olds. "I further believe that eventually this plan will prove to be bad for this country. The automobile industry has become of such tremendous importance and such gigantic proportions, that anything that affects it affects the country as a whole."

Although some expensive items, such as pianos and sewing machines, had been sold on time before 1920, it was time sales of automobiles that set the precedent during the twenties for a great extension of consumer installment credit. By 1926 time sales accounted for about three-fourths of all automobile sales. And as the market for automobiles approached saturation after 1925, the finance companies, fearing they had overextended credit for automobile purchases and wishing to diversify their risks, played an active role in encouraging installment purchases of many other types of merchandise. The Hoover Committee on

Recent Economic Changes reported in 1929 that "simultaneously with the advance in the use of the automobile there has been a marked advance in the purchase of many commodities that a decade ago would have been described as luxury goods, but which have since entered so universally into the average budget as no longer to be regarded as such." A 1926 survey made by the Eberle and Riggleman Economic Service of Los Angeles showed that the automobile installment commitments of California car buyers averaged 18 percent of monthly income. Walter Engard, a Warren, Ohio, automobile dealer, gave a typical simple-minded assessment of how time sales supposedly benefitted America's society and economy. "Higher standards of living are built up through the millions of individual extravagances," wrote Engard in *Motor*. "To keep America growing we must keep Americans working, and to keep Americans working we must keep them wanting; wanting more than the bare necessities; wanting the luxuries and frills that make life so much more worthwhile, and installment selling makes it easier to keep Americans wanting."

John C. Burnham, the leading authority on the gasoline tax, is amazed at the extent to which "Americans were willing to pay for the almost infinite expansion of their automobility." Public support for heavy motor vehicle special use taxes is a case in point. Motorists early came to support higher and annual registration fees as one means of securing better roads. For the same reason, there has consistently been almost no public opposition to the gasoline tax. The gasoline tax was innovated in Oregon, New Mexico, and Colorado in 1919 as a means of raising money to meet matching funds for road construction that the federal government made available to the states after World War I. By 1925 gasoline taxes had been imposed by forty-four states and the District of Columbia. By 1929 all states collected gasoline taxes, which amounted to some $431 million in revenue that year, and rates of three and four cents a gallon were common. In 1921 road construction and maintenance were financed mainly by property taxes and general funds, with only about 25 percent of the money for roads coming from automobile registration fees. By 1929 gasoline taxes were the main source of revenue for highway expenditures, and twenty-one states no longer used any general property taxes for main roads. The reasoning was that "the gasoline tax was superior as a user tax because the amount of gasoline consumed in a vehicle was a good measure of the use of the road and also of the damage that a vehicle did to a road. . . . the tax was 'equitable' in itself and also that those who paid it benefited directly." The chief collector of the gasoline tax in Tennessee exclaimed in 1926, "Who ever heard, before, of a popular tax?" And Burnham has pointed out that "never before in the history of taxation has a major tax been so generally accepted in so short a period."

"Among the many factors which contributed toward the expansion of local taxes [between 1913 and 1930], probably no single one, price inflation aside, exercised a more potent influence than did the automobile," reported the President's Research Committee on Social Trends in 1933. "This taking to wheels of an entire population had a profound effect on the aggregate burden of taxation." Staggering highway expenditures, the bulk of which came directly out of the motorists' pockets in use taxes, accounted for only part of this increased tax burden. The committee found that "it was not merely in its influence on highway costs that the automobile affected the size of the tax bill. Its use in cities created serious problems of traffic congestion and increased crime. Motorized police and traffic control became important items of increased expenditure. Moreover, . . . the motorcar was responsible for a spreading out of the population of cities toward the peripheral or suburban areas. This movement helped to swell the volume of local taxation, since schools and other public facilities had to be provided anew in these outer areas, despite the under-utilization of such facilities in older areas where population declined. In the rural sections of the country, good roads and the motor bus stimulated the growth of the consolidated school, thus putting within the reach of farm children an educational offering which approached the urban standard. In terms of educational returns per dollar spent, it was much more economical than the little one-room school. Nevertheless, it involved a larger absolute amount of expenditure and rural tax rates rose accordingly." The committee concluded that "since the changes which came with the motor era are inextricably bound up with other types of change, it is impossible to state in dollars and cents just how much the automobile has cost the taxpayers of the country."

This was indeed an ironic outcome from the adoption of an innovation whose proponents had claimed, less than a generation before, would lower the cost of living. Thomas L. White had spoken the mind of the public as well as of the experts when he predicted in 1910 that by relieving "the situation arising out of the diminished fertility of the soil, the depopulation of the rural districts, the growing congestion of the cities, and the decreasing food value of the dollar . . . the self-propelled vehicle will justify from a national standpoint the expenditure of brains and money hitherto so freely lavished on its development." But by the late 1920s our historical experience was proving that these and most other early predictions about the benefits of mass personal automobility were erroneous.

The President's Research Committee on Social Trends noted that "car ownership has created an 'automobile psychology'; the automobile has become a dominant influence in the life of the individual and he, in a very real sense, has become dependent on it." Blaine A. Brownell has recently made a well-phrased assessment of the motorcar's cultural meaning in the 1920s. "It was clear to almost all observers in the 1920s that technology was a highly significant factor in altering the past and shaping the future," he writes. "Whether technology was to be welcomed as symbolic of a new and more prosperous age, or damned as subversive of age-old values and mores is perhaps less important than the fact that all observers tended to see technology as a fundamental force in American life, and the automobile as probably the most significant of the technological innovations appearing in the period. The motor vehicle was a more impressive piece of machinery than a radio, more personal in its impact than a skyscraper or dynamo, and certainly more tangible than electricity. Thus it was generally more legible as a symbol and more apparent in its consequences."

Brownell's work on the automobile in southern urban areas in the 1920s stands as a model of what needs to be done for American culture as a whole. Even though the South lagged behind the rest of the nation in adopting the automobile, "the motorcar's overall influence on the South was massive. The region's transportation system was probably revolutionized to a greater extent by the motor vehicle than was the case elsewhere, and the traditional provinciality of the rural South was radically altered by new highways." By 1929 the percentage of retail businesses listed in the automotive category by the Bureau of the Census for major southern urban areas ranged from 14.2 in New Orleans to 20.7 in Birmingham, Alabama. "Its total economic significance is virtually impossible to compute with precision, but it would probably be measured in the billions of dollars in major southern cities alone." Sunday blue laws gave way to automobility, and problems caused by the automobile became the most time-consuming item on the agendas of southern city councils. As the historian Thomas D. Clark observed in 1961, "Detroit automobile makers set off the most effective Yankee invasion that ever disturbed southern complacency," an invasion that "had more long-range economic meaning for the South than all the Civil War generals combined. The established way of life in the South was shaken to its very foundation by this new Yankee machine."

"Highly favorable reactions to the automobile were frequently voiced in the major daily newspapers in southern cities," finds Brownell. "General community sentiment, to the extent that it was mirrored in these editorial columns, was that the motorcar contributed to 'progress' and to the prospects for material prosperity. An era marked by widespread automobile travel was welcomed as one both modern and affluent." He notes further that "the opinions of all groups toward the motor vehicle are, in the final analysis, impossible to fully determine, largely because most such opinions were never preserved in the historical record. . . . and because many black and labor union publications were at best but partially representative of the sentiments prevailing in either the black community or among white workingmen. On the basis of the existing evidence, however, attitudes toward the automobile cannot be differentiated along racial or class lines." Among the groups for which adequate data exist, "the motor vehicle was especially lauded by the southern urban business community, largely because it promised to open up new channels of commerce, expand the pool of customers for downtown merchants, and make available large expanses of outlying territory for urban growth and economic development."

A major contribution of Brownell's work is that he demonstrates for the first time that as the 1920s wore on, many problems of mass personal automobility began to become apparent. He believes that "the notion advanced by some historians that the motor vehicle was accepted uncritically and with little awareness of its potential consequences is a false one." According to Brownell, the seeds of our contemporary discontent with the motorcar were already beginning to germinate by the late 1920s. "Then as now, ownership of automobiles did not necessarily reflect a complete and unquestioning acceptance of their consequences. The views expressed about the motor vehicle in the 'circulating media' in southern cities during the decade . . . suggest a much more complex pattern of response. Such views ran the gamut from complete acceptance of the automobile to rather deep suspicions concerning its impact on American life. Not surprisingly, most conceptions of this significant technological innovation were highly ambiguous. The vehicle that promised to infinitely expand the radius of individual mobility also seemed to threaten the tightly knit family unit and prevailing moral standards; the same automobile that was supposed to decentralize the city and improve urban access to the countryside acted paradoxically to render the city even more congested; and the motor vehicle that epitomized freedom from restraint was to impose new restrictions on the possibilities and style of urban living." Although atmospheric pollution by the motor vehicle was not yet recognized as a problem, "as the decade progressed many business-oriented publications and organizations expressed a rising concern about the motorcar's threat to streetcar lines and the growing toll of deaths and injuries for which it was responsible. Downtown merchants were especially worried over the motor traffic congestion that plagued most central cities throughout the period. . . . Many writers, especially clergymen, lamented the auto's allegedly unfavorable impact on traditional moral standards and the family unit, though most of these observations were apparently jeremiads aimed at undesirable secular forces that seemed to be undermining older values—and which the automobile seemed to symbolize perfectly."

The most comprehensive contemporaneous analysis of the automobile culture in the 1920s and the 1930s is in the Lynds's Middletown series. "As, at the turn of the century, business class people began to feel apologetic if they did not have a telephone, so ownership of an automobile has now reached the point of being an accepted essential of normal living," the Lynds wrote in 1929. They found in the depths of the Great Depression, however, that "if the word 'auto' was writ large across Middletown's life in 1925, this was even more apparent in 1935, after six years of depression. One was immediately struck in walking the streets by the fact that filling stations have become in ten years one of the most prominent physical landmarks. . . . In 1925 Middletown youngsters, driven from street play to the sidewalks, were protesting, 'Where can I play?' but in 1935 they were retreating even from the sidewalks, and an editorial, headed 'Sidewalk Play is Dangerous,' said, 'It is safe to say that children under the age of eight years should not be permitted to play on sidewalks.' . . . While some workers lost their cars in the depression, the local sentiment, as heard over and over again, is that 'People give up everything in the world but their cars.' According to a local banker, 'The depression hasn't changed materially the value Middletown people set on home ownership, but *that's* not their primary desire, as the automobile always comes first.'" It was apparent to the Lynds in 1937 that, "to a considerably greater extent than in 1925, Middletown's life is today derived from the automotive industry—and the city is aware of it to its marrow!"

During the depression the Middletown car culture if anything had more symbolic meaning to the working class than to the business class. "If the automobile is by now a habit with the business class, a comfortable, convenient, pleasant addition to the paraphernalia of living, it represents far more than this to the working class; for to the latter it gives the status which his job increasingly denies, and, more than any other possession or facility to which he has access, it symbolizes living, having a good time, the thing that keeps you working." Indeed, automobility was probably the main barrier to the development of class consciousness among workers in the 1920s and 1930s. "Working in an open-shop city with its public opinion set by the business class, and fascinated [even in the depression!] by a rising standard of living offered them on every hand," noted the Lynds, "[workers] do not readily segregate themselves from the rest of the city. They want what Middletown wants, so long as it gives them their great symbol of advancement—an automobile. Car ownership stands to them for a large share of the 'American dream'; they cling to it as they cling to self-respect, and it was not unusual to see a family drive up to a relief commissary in 1935 to stand in line for its four- or five-dollar weekly food dole. 'It's easy to see why our workers don't think much about joining unions,' remarked a union official in 1935. He went on to use almost the same words heard so often in 1925: 'So long as they have a car and can borrow or steal a gallon of gas, they'll ride around and pay no attention to labor organization; and if they can't get gas, they're busy trying to figure out some way to get it.'"

Automobility was undoubtedly the major force unifying Americans in the period between the two world wars. Despite comments in the automobile trade journals in the early 1920s recognizing that "illiterate, immigrant, Negro and other families" were "obviously outside" the market for motorcars, there was agreement that "every clear-thinking American, be he rich or poor, realizes that in a measure, the automobile and its manufacturers are helping to solve the labor and social problems of the future," as well as "the tendency of the automobile to bring into intimate and helpful contact sections of our population which normally would never meet." The automobile outing and the automobile vacation became national institutions in the 1920s. By 1926 some 5,362 "motor camps" dotted the American countryside, and an avalanche of tourists who never before had traveled more than a few miles from home began to descend on distant national parks, forests, and points of historic interest. The new mobility of the labor force and long-distance trucking decentralized the location of industrial plants, opened up the Pacific Coast and the Southwest to commercial development, knit regional economies more tightly together, and promised eventually to abolish class and ethnic, as well as sectional, differences. "Scarcely ever before have rich and poor, educated and ignorant, self-styled superiors and acknowledged inferiors, landlords and landless, white folks and black, ridden in the same type of vehicle; and never before have they ridden so fast that they could not see who was approaching," said Arthur Raper in a 1936 study of two rural Georgia counties. "The opportunities afforded by the automobile provide a basis for a new mobility for whites as well as Negroes, based upon personal standards rather than upon community mores—upon what the individual wants to do rather than what the community does not want him to do."

There was ample evidence by the 1920s that the national, individualized automobile culture was destructive of the beneficial as well as the repressive aspects of "community." "The mobility afforded by new modes of transportation combined with . . . periodic waves of employment, unemployment, and reemployment to diminish the tendency for the workers in a given factory to live together immediately about the plant," observed the Lynds. "The trend towards decentralization of workers' dwellings means that instead of a family's activities in getting a living, making a home, play, church-going, and so on largely overlapping and bolstering each other, one's neighbors may work at shops at the other end of the city, while those with whom one works may have their homes and other interests anywhere from one to two-score miles distant." The breakdown of the neighborhood was also evident in that "the housewife with leisure does not sit so much on the front porch in the afternoon after she 'gets dressed up,' sewing and 'visiting' and comparing her yard with her neighbors', nor do the family and neighbors spend long summer evenings and Sunday afternoons on the porch or in the side yard since the advent of the automobile and the movies." A Middletown housewife explained, "In the nineties we were all much more together. People brought chairs and cushions out of the house and sat on the lawns evenings. We rolled out a strip of carpet and put cushions on the porch step to take care of the unlimited overflow of neighbors that dropped by. We'd sit out so all evening. The younger couples perhaps would wander off for half an hour to get a soda but come back to join in the informal singing or listen while somebody strummed a mandolin or guitar."

E. C. Stokes, a former governor of New Jersey and the president of a Trenton, New Jersey, bank, claimed in 1921: "Next to the church there is no factor in American life that does so much for the morals of the public as does the automobile. . . . Any device that brings the family together as a unit in their pursuit of pleasure is a promoter of good morals and yields a beneficent influence that makes for the good of American civilization. If every family in the land possessed an automobile, family ties would be closer and many of the problems of social unrest would be happily resolved. . . . The automobile is one of the country's best ministers and best preachers."

Contrary to Stokes's widely shared expectations, by 1929 it was evident to the Lynds and many others that any tendency of automobility to bring the family together was "a passing phase"—although an increasing number of people did tend to find the Sunday drive preferable to attending church. "No one questions the use of the auto for transporting groceries, getting to one's place of work or the golf course, or in place of the porch for 'cooling off after supper,'" the Lynds found. "But when auto riding tends to replace the traditional call in the family parlor as a way of approach between the unmarried, 'the home is endangered,' and all-day Sunday motor trips are a 'threat against the church'; it is in the activities concerned with the home and religion that the automobile occasions the greatest emotional conflicts." Although no one has ever proved that the back seat of a Model T was more convenient or comfortable than a haystack, of thirty girls brought before the Middletown juvenile court for so-called "sex crimes," nineteen named an automobile as the place where the offense occurred, and a Middletown judge explained a dwindling red-light district by pointing out that "the automobile has become a house of prostitution on wheels." "What on earth *do* you want me to do? Just sit around home all evening!" retorted a popular high school girl . . . when her father discouraged her going out motoring for the evening with a young blade in a rakish car waiting at the curb."

John B. Rae may well be right in his claim that such criticisms of the automobile simply "exemplified man's propensity to blame his technology rather than himself for whatever evil consequences it might produce." Yet Rae dismisses too facilely the fact that automobility undermined the family as an institution. Although in theory the family car could bring husbands, wives, and children together in their leisure-time activities, the divorce rate continued to climb in the 1920s, and intergenerational conflicts between parents and children reached a new height during the decade. There is no evidence that the motorcar contributed to the divorce rate, but then neither did it, as early proponents of automobility expected, stop the divorce mill from grinding. That the motorcar undercut parental authority, on the other hand, is unequivocal. The Lynds pointed out, for example; "The extensive use of this new tool by the young has enormously extended their mobility and range of alternatives before them; joining a crowd motoring over to a dance in a town twenty miles away may be a matter of a moment's decision, with no one's permission asked. Furthermore, among the high school set, ownership of a car by one's family has become an important criterion of social fitness: a boy almost never takes a girl to a dance except in a car; there are persistent rumors of the buying of a car by local families to help their children's social standing in high school."

Mounting automobile registrations are undoubtedly the best indication that Americans in the 1920s continued to view automobility on balance as highly beneficial. Automobile sales dropped sharply in the Great Depression. "Car ownership in Middletown," nevertheless, "was one of the most depression-proof elements of the city's life in the years following 1929—far less vulnerable, apparently, than marriages, divorces, new babies, clothing, jewelry, and most other measurable things both large and small. . . . The passenger-car registrations in Middletown's entire county not only scarcely registered any loss in the early years of the depression but, both in numbers and in ratio to population, stood in each of the years 1932-35 above the 1929 level. Along with this tough resistance of Middletown's habit of car owning to the depression undertow, went a drop of only 4 percent in the dollar volume of gasoline sales in Middletown between 1929 and 1933." For the nation as a whole, motor vehicle miles of travel increased from 198 billion in 1929 to 206 billion in 1930 and 216 billion in 1931. Their depression low of 201 billion occurred in both 1932 and 1933. Special motor vehicle tax receipts progressively mounted through the depression from $849 million in 1929 to $990 million in 1932, the trough of the depression, reaching $1.693 billion in 1940. Highway expenditures were significantly greater throughout the depression than they had been in the supposedly prosperous 1920s. Motor vehicle registrations decreased from 26.75 million in 1930 to a low of 24.159 million in 1933, then bounced back to 32.453

million in 1940. "While, therefore, people were riding in progressively older cars as the depression wore on, they manifestly continued to ride."

Other data, however, indicate that for the bulk of the population the benefits of automobility, even by the early 1920s, were probably mainly symbolic. A questionnaire mailed by the National Automobile Chamber of Commerce in 1920 to a random sample of automobile owners in "ten widely selected states" showed scant evidence that automobility was improving the quality of car owners' lives. Of the owners who responded, 90 percent said their cars were "used more or less for business," and 60 percent of the mileage driven was "for business purposes." The NACC estimated a 57 percent "gain through car use over previous income or efficiency." The rank order of gains reported, by occupation, was real estate and insurance agents (113 percent), medical doctors (104 percent), salesmen (103 percent), clergymen (98 percent), school supervisors (72 percent), farmers (68 percent), contractors (51 percent), and manufacturers and bankers (33 percent). The most significant point was that only 37 percent of the owners reported "improving living conditions through use of the car (suburban life, etc.)," meaning that almost two-thirds of the automobile owners who responded found that the quality of their lives either had remained about the same or had deteriorated with automobility. There was thus much more ironic truth than its author could have recognized in a 1923 quip in *Motor*: "It may be that Kipling is right, and that we have lost our national soul if we ever had one. We may be a nation of mechanical, money-grubbing 'robots.' But certainly the automobile touched either a soul or some controlling springs and gears in our national life."

The main things that automobility symbolized were material prosperity through a higher standard of living, individual mobility, and an improvement in the quality of life through a fusing of rural and urban advantages. If one excludes the obviously great payoff for such economic interest groups as the automobile and oil industries, real estate developers, and contractors, the realities of the automobile revolution fell far short of its proponents' promises for most Americans. As we have already seen, what the individual gained from automobile ownership was at the expense of undermining community and family, and it invited anonymity and anomie. Improvements in the quality of life through a fusing of rural and urban advantages also proved illusory.

Reynold M. Wik's "grass-roots Americans" undoubtedly benefitted more than the urban masses from personal automobility. The motorcar and improved roads increased rural land values, lessened the drudgery of farm labor, ended rural isolation, and brought the farm family the amenities of city life, the most important being far better medical care and schools. In a day when "busing" has come to be associated in the minds of many Americans with an alleged lowering of educational standards, it seems particularly pertinent to point out that in the 1920s and 1930s the daily busing of farm children long distances to consolidated schools was hailed as a significant forward step in achieving educational parity between rural and urban schools.

These benefits of the school bus and the Model T notwithstanding, the cityward migration of farm youth was not stemmed, as auto enthusiasts promised it would be. Even more important, the family farm was already being killed off in the 1920s by a combination of the farm tractor, corporate agriculture, and the propensity of bankers to foreclose mortgages that had been incurred to produce the farmers' automobility. One of the main symbols of the dispossession of rural Americans in the 1930s, chronicled so eloquently in John Steinbeck's *The Grapes of Wrath*, was "the machine man driving a dead tractor on land he does not know and love . . . contemptuous of the land and of himself." And in Steinbeck's novel, automobility ended up as the frustrating search for daily subsistence of the migrant agricultural worker.

Proponents of automobility were right about the necessity of banishing the unsanitary horse from cities, and they were right that a viable urban transportation system could not run entirely on rails but also required great dependence on the more flexible motor vehicle. From the perspective of rational hindsight, however, the best solution to urban transportation problems would have been a balanced commitment to motortrucks and motor buses, along with rail transportation, rather than the mass ownership of private passenger cars in cities. By the 1920s it was obvious, contrary to the predictions of early proponents of automobility, that mass dependence on passenger cars had compounded urban traffic congestion and parking problems. Although it was not yet recognized, the urban neighborhood as a community was already being destroyed by the decentralization and segregation of activities encouraged by automobility and by the longer blocks, combined with widened streets and narrower sidewalks, that accommodation to the motorcar demanded. Nor was it foreseen that auto exhaust in the antiseptic, horseless city was going to prove even more detrimental to public health than horse exhaust once had been.

As early as 1913 a spokesman for the Merchants Association of New York complained to that city's Motor Truck Club at its October 15 meeting: "The growth of the automobile industry and automobile traffic has gone on faster than our ability to care for it properly. The automobile manufacturer and dealer has been so busy selling automobiles that he hasn't even taken the time to give his customers proper treatment or his broken-down machines the proper kind of convalescent care, let alone giving attention to traffic conditions, which are primarily municipal and interest him only indirectly." By the end of 1916 even the editors of *Automobile* were overwhelmed that "every day in big cities the parking problem grows more acute. If it is bad today, and indeed it is so, what will be the situation in 3 years? We are facing something which was never foreseen in the planning of our towns, a thing which has come upon us so swiftly that there has been no time to grasp the immensity of the problem till we are almost overcome by it." The situation had become so grave by 1920 that *Motor Age* felt that "the collapsible wheel-base seems to be about the only solution of the traffic problem as it exists in our larger cities today. There is an alleged joke in vaudeville about the manufacturer of a certain well known car [Ford] who is considering putting wheels on the rear end of his cars so that he can run them on end and so get more of them on the streets. But if you watch the stream of motor traffic in any of our large cities at the 'rush hours,' you will realize that that joke isn't so all-fired funny after all." By 1923 the automobile trade journals considered the urban traffic problem "one of the great problems of the day." Articles in *Motor*, for example, warned, "Stop! You Are Congesting the Streets," and asked, "Will Passenger Cars Be Barred from City Streets?"

"The only means of affording substantial relief from this burden [of urban traffic problems] as far as most city leaders were concerned was city planning—a concept that was advanced as a virtual panacea for a whole range of urban ills, but was always fundamentally tied to the demands posed by motor vehicles," Brownell concludes. "The ultimate failure to significantly ease the impact of the automobile occurred even though the responses of city governments and local leaders to the automotive challenge was in the best American pragmatic tradition. As the numbers of automobiles mounted, so did the governmental response: new taxes, improved roads, expanded parking facilities, extensive surveys, and a vast system of regulations enacted to guarantee the auto's operation in the public interest and welfare." Thus, instead of attempting to discourage the use of private passenger cars in cities, politicians and city planners adopted the expensive and ultimately unworkable policy of unlimited accommodation to the motorcar. That American urban life would conform to the needs of automobility rather than vice versa was obvious by the early 1920s. In 1923, for example, the Chicago Automobile Trade Association was shown plans drafted by the Chicago Plan Commission for "vast projects . . . to widen streets and extend boulevards, open more through streets, eliminate jogs, straighten the Chicago river, and connect up the whole street system so that automobile traffic may be conducted with convenience." The association's president thought it "imperative that something be done to make room for the hundreds of automobiles being put on the streets by the dealers each week." And he stressed that "of all the business interests of the city the automobile dealers should be most active in . . . support [of the Chicago Plan Commission] because of what it means to their business."

Some 135,000 suburban homes in sixty cities were already chiefly dependent on motor vehicle transportation by 1922, and the exodus of the middle class and businesses accelerated as the automobile culture matured in the 1920s. Massive conversion of the central city to accommodate the motorcar was thus paradoxically undertaken as its cost and problems were avoided by the escape to suburbia that automobility encouraged. *Motor* enthused that, in making possible the proliferation of suburban country clubs, the motorcar had "added a new phase of growing social importance to the social life of this country." But a more significant development in American social life was that the central city increasingly came to be an isolated ghetto of the deprived of American society as the middle class retreated to a modified version of the Jeffersonian vision of a nation of yeoman farmers. The implications of this progressive bifurcation of American society into urban ghettos and middle-class residential suburbs would not become apparent to the middle class, unfortunately, until the deterioration of living conditions in the central city manifested itself in widespread riots during the mid-1960s.

Automotive safety, on the other hand, was already recognized as a major problem by the mid-1920s. Although improved roads and better control of motor vehicle traffic progressively reduced the number of fatalities per miles of automobile travel in the 1920s, more automobiles traveling at higher speeds meant a mounting toll of fatalities, injuries, and property damage in absolute figures. In 1924, automobile accidents accounted for 23,600 deaths (including the deaths of 10,000 children), over 700,000 injuries, and over a billion dollars in property damage— "regarded as one of the big economic problems of the day" by *Motor Age.*

The automobile industry consistently turned out cars with more horsepower that were capable of higher speeds than could be driven legally, much less safely, on the highways of the day. The industry also put styling ahead of safety in automotive design. Newspaper accounts of accidents frequently mentioned the failure of steering mechanisms, brakes, tires, and other components. Yet the NACC Traffic and Safety Committee analysis of 280 automobile accident fatalities in 1924 traced only 7 to a defect in the vehicle. According to the NACC committee, physical conditions, such as visibility and conditions of the pavement, accounted for 92 of the fatalities, while 150 were the fault of pedestrians. The remaining 152 were attributed to the motorist, with excessive speed accounting for 48 and violations of "rules of the road" for 40. To reduce accidents the NACC committee recommended depriving "the careless motorist" of his car, giving better instruction on how to drive in wet weather, adjusting headlights properly, guarding children better, appointing traffic administrators, and keeping records of the causes of fatalities.

The automobile industry dominated the National Safety Council, and about all that came out of the council's 1925 and 1926 meetings was a warm endorsement of uniform state traffic codes and the pious statement that "more must be done to see that those using the highways come to have a better understanding of what the laws are and additional educational work must be carried on to make the public realize the urgent need of being more careful. Several speakers expressed the opinion that there are thousands holding operators' licenses who are not fit to drive cars. Some expressed the opinion that applicants for operators' licenses should be compelled to undergo at least an optical, hearing, and mental test if not a general physical examination." The industry prided itself, however, that NACC statistics showed that "the traffic accident situation isn't as bad as some make it."

It was evident in the specialized automobile journals that automobility had stopped being a progressive force for change and by the mid-1920s was beginning to petrify into a complacent maturity. The journals had been inaugurated as heralds of a new horseless age. Most tried to cover the entire spectrum of the automobile movement and were addressed to the motoring public. By World War I, however, they had become narrow trade journals catering to the automobile industry, especially to the interests of automobile dealers. *Motor* held out the longest, but in 1924, after twenty-one years of publication for car owners, it, too, became a trade magazine. By then the earlier confident predictions about the utopian benefits of mass automobility in *Motor*'s columns had given way to superficial platitudes. "Motoring is one of the most effective forms of health insurance that modern life has produced, as a palliative to its intensified, speeded-up social and commercial processes," said the health commissioner of New York City in a 1922 *Motor* article. "Do you realize, that without the motorcar golf could never have become the popular game that it is today." "An automobile is as much an asset to a present-day American citizen as a deposit in a bank," said another writer. "There has been a tendency in some quarters to an expression of opinion that the automobile industry has about reached the limit of its development. . . . that the industry has already passed this limit and has persuaded the country to absorb more cars than it was justified in carrying." He argued, however, that "this country has got to maintain its present ratio of motor vehicles, because its life has been gradually formed upon the foundations of speed and mobility that the motor vehicle has brought to it." *Motor* asked Thomas A. Edison in 1924: "If the automobile develops in the next twenty-five years at the same rate as in the past twenty-five, what will be the most startling transformation in everyday life?" The best answer that the dean of American inventors could give was: "Everybody who can will go out camping in the summer." . . .

CHAPTER 9

Crabgrass Frontier: The Suburbanization of the United States

Kenneth T. Jackson

By combining the best elements of the city and the countryside, the suburban ideal has been enormously attractive to many, and not just in modern times. A letter to the king of Persia (written in cuneiform on a clay tablet) in 539 B.C. contained these remarks: "Our property seems to me the most beautiful in the world. It is so close to Babylon that we enjoy all the advantages of the city, and yet when we come home we are away from all the noise and dust." In North America, distinct suburbs were established in Boston, Philadelphia, and New York well before the Revolutionary War. But it was in the post–World War II period that American suburbs expanded to an extent and at a tempo that was historically unprecedented. Our selection from Kenneth T. Jackson's *Crabgrass Frontier* provides a brief survey of this phenomenon, and intertwines discussion of the techniques of mass housing construction with related social and economic trends in the postwar period. Of all the many forms of modern technology that influence our daily lives, residential architecture is one of the most important — and most overlooked.

We begin by looking at the pent-up demand for housing that began to grow during World War II. Marriage and birth rates climbed, then soared at the end of the war, at a time when there were virtually no homes for sale or apartments for rent. Many of the returned veterans and their families were doubled up with relatives and friends. The construction industry responded, but Jackson points out the critical role played by the federal government, which underwrote construction with mortgage insurance. It was national policy that veterans should have homes of their own.

Residential construction, until this time, was undertaken almost entirely by house builders who operated on a small scale and built few homes in any given year. But one family firm on Long Island, Levitt and Sons, revolutionized the home-building industry by bringing the techniques of mass construction to suburban development. The Levitts turned 4,000 acres of potato farms on Long Island into Levittown, the site of ultimately 17,000 homes, the biggest private housing project in American history. A second Levittown would follow in Pennsylvania, and a third in New Jersey.

Jackson explains that the Levitts departed from a number of existing conventions in the construction industry, such as their use of preassembled components wherever possible. The typical Cape Cod model was a small (750 square feet) two-bedroom model, which Jackson, with understatement, describes as "unpretentious." Architectural critics deplored the repetition of identical homes on identical lots. But the house's most important feature was its affordable price, and Levittown turned the detached, single-family home into a real possibility for the white middle class. It was a hugely successful commercial success.

Other builders soon used many of the Levitt methods to build sprawling suburban developments in other metropolitan areas. The new mass construction technology could not have been used, however, without the federal housing programs that made it possible for builders to arrange the financing of such large-scale projects.

These postwar suburbs shared certain characteristics that had social implications, Jackson observes. The new suburbs were located on the periphery, either outside of existing cities or, if inside cities, in the outermost districts. They consisted of single, detached homes on separate lots, a low-density design that assumed that residents would have their own automobiles. They were architecturally similar, contributing to the disappearance of distinctive regional styles in American architecture. Finally, these suburbs brought home ownership within the range of families who had never been able to afford their own homes; yet strict racial discrimination kept their democratic potential from fuller realization.

The suburbs' residents were a homogenous group: The same race, the same class, the same age, all living in the same homes. As early as the 1950s, critics worried that the suburbs were breeding mindlessly conformist Organization Men. Here we encounter interesting questions about the relationship between modern technology and our inner nature: Do we tend to become identical to one another if we live in identical homes situated on identical streets?

Measured by technical improvements, the contractor-built homes in the postwar suburbs marked significant advances over previous standards. The homes had central heating, indoor plumbing, refrigerators, and other amenities. But Jackson argues that these material improvements were accompanied by other kinds of changes that meant losses, not gains, for American life. The extended family was weakened as young families moved into houses of their own, and housewives were isolated from the stimulation of work and from contact with employed adults. Lewis Mumford mused that the suburb, which had originally been built to serve as an escape from the city, offered no escape from its own numbing conformity.

To open this reading selection, Jackson provides a quotation from William Levitt, the Long Island builder: "No man who owns his own house and lot can be a Communist. He has too much to do." It's a funny line, and it also can be read somewhat seriously too. Maybe the suburbs did contribute to political quiescence in postwar America. Today, housing problems are claiming more and more attention, with numbers of homeless people visible in many cities, and with millions of young working couples finding themselves priced out of many housing markets. One wonders whether the dream of the single, detached home will fade as the possibility of its realization grows more remote. Will the next generation be able eventually to afford homes equal to those of the Levittown of the 1940s?

THE BABY BOOM AND THE
AGE OF THE SUBDIVISION

What the Blandings wanted . . . was simple enough: a two-story house in quiet, modern good taste, . . . a good-sized living room with a fire place, a dining room, pantry, and kitchen, a small lavatory, four bedrooms and accompanying baths . . . a roomy cellar . . . plenty of closets.

—Eric Hodgins, *Mr. Blandings Builds His Dream House* (1939)

No man who owns his own house and lot can be a Communist. He has too much to do.

—William J. Levitt, 1948

At 7 P.M. (Eastern time) on August 14, 1945, radio stations across the nation interrupted normal programming for President Harry S. Truman's announcement of the surrender of Japan. It was a moment in

---

SOURCE: From *The Crabgrass Frontier: The Suburbanization of the United States* by Kenneth T. Jackson. Copyright © 1985 by Oxford University Press, Inc. Reprinted by permission.

time that those who experienced it will never forget. World War II was over. Across the nation, Americans gathered to celebrate their victory. In New York City two million people converged on Times Square as though it were New Year's Eve. In smaller cities and towns, the response was no less tumultuous, as spontaneous cheers, horns, sirens, and church bells telegraphed the news to every household and hamlet, convincing even small children that it was a very special day. To the average person, the most important consequence of victory was not the end of shortages, not the restructuring of international boundaries or reparations payments or big power politics, but the survival of husbands and sons. Some women regretted that their first decent-paying, responsible jobs would be taken away by returning veterans. Most, however, felt a collective sigh of relief. Normal family life could resume. The long vigil was over. Their men would be coming home.

In truth, the United States was no better prepared for peace than it had been for war when the German *Wehrmacht* crossed the Polish frontier in the predawn hours of September 1, 1939. For more than five years military necessity had taken priority over consumer goods, and by 1945 almost everyone had a long list of unfilled material wants.

Housing was the area of most pressing need. Through sixteen years of depression and war, the residential construction industry had been dormant, with new home starts averaging less than 100,000 per year. Almost one million people had migrated to defense areas in the early 1940s, but new housing for them was designated as "temporary," in part as an economy move and in part because the real-estate lobby did not want emergency housing converted to permanent use after the war. Meanwhile, the marriage rate, after a decade of decline, had begun a steep rise in 1940, as war became increasingly likely and the possibility of separation added a spur to decision-making. In addition, married servicemen received an additional fifty dollars per month allotment, which went directly to the wives. Soon thereafter, the birth rate began to climb, reaching 22 per 1,000 in 1943, the highest in two decades. Many of the newcomers were "good-bye babies," conceived just before the husbands shipped out, partly because of an absence of birth control, partly because the wife's allotment check would be increased with each child, and partly as a tangible reminder of a father who could not know when, or if, he would return. During the war, government and industry both played up the suburban house to the families of absent servicemen, and between 1941 and 1946 some of the nation's most promising architects published their "dream houses" in a series in the *Ladies' Home Journal*.

After the war, both the marriage and the birth rates continued at a high level. In individual terms, this rise in family formation coupled with the decline in housing starts meant that there were virtually no homes for sale or apartments for rent at war's end. Continuing a trend begun during the Great Depression, six million families were doubling up with relatives or friends by 1947, and another 500,000 were occupying quonset huts or temporary quarters. Neither figure included families living in substandard dwellings or those in desperate need of more room. In Chicago, 250 former trolley cars were sold as homes. In New York City a newly wed couple set up housekeeping for two days in a department store window in hopes that the publicity would help them find an apartment. In Omaha a newspaper advertisement proposed: "Big Ice Box, 7 × 17 feet, could be fixed up to live in." In Atlanta the city bought 100 trailers for veterans. In North Dakota surplus grain bins were turned into apartments. In brief, the demand for housing was unprecedented.

The federal government responded to an immediate need for five million new homes by underwriting a vast new construction program. In the decade after the war Congress regularly approved billions of dollars worth of additional mortgage insurance for the Federal Housing Administration. Even more important was the Servicemen's Readjustment Act of 1944, which created a Veterans Administration mortgage program similar to that of FHA. This law gave official endorsement and support to the view that the 16 million GI's of World War II should return to civilian life with a home of their own. Also, it accepted the builders' contention that they needed an end to government controls but not to government insurance on their investments in residential construction. According to novelist John Keats, "The real estate boys read the Bill, looked at one another in happy amazement, and the dry, rasping noise they made rubbing their hands together could have been heard as far away as Tawi Tawi."

It is not recorded how far the noise carried, but anyone in the residential construction business had ample reason to rub their hands. The assurance of federal mortgage guarantees—at whatever price the builder set—stimulated an unprecedented building boom. Single-family housing starts spurted from only 114,000 in 1944, to 937,000 in 1946, to 1,183,000 in 1948, and to 1,692,000 in 1950, an all-time high. However, as Barry Checkoway has noted, what distinguished the period was an increase in the number, importance, and size of large builders. Residential construction in the United States had always been highly fragmented in comparison with other industries, and dominated by small and poorly organized house builders who had to subcontract much of the work because their low volume did not justify the hiring of all the craftsmen needed to put up a dwelling. In housing, as in other areas of the economy, World War II was beneficial to large businesses. Whereas before 1945, the typical contractor had put up fewer than five houses per year, by 1959, the median single-family builder put up twenty-two structures. As early as 1949, fully 70 percent of new homes were constructed by only 10 percent of the firms (a percentage that would remain roughly stable for the next three decades), and by 1955 subdivisions accounted for more than three-quarters of all new housing in metropolitan areas.

Viewed from an international perspective, however, the building of homes in the United States remained a small-scale enterprise. In 1969, for example, the percentage of all new units built by builders of more than 500 units per year was only 8.1 percent in the United States, compared with 24 percent in Great Britain and 33 percent in France. World War II, therefore, did not transform the American housing industry as radically as it did that of Europe.

Levittown

The family that had the greatest impact on postwar housing in the United States was Abraham Levitt and his sons, William and Alfred, who ultimately built more than 140,000 houses and turned a cottage industry into a major manufacturing process. They began on a small scale on Long Island in 1929 and concentrated for years on substantial houses in Rockville Center. Increasing their pace in 1934 with a 200-unit subdivision called "Strathmore" in Manhasset, the Levitts continued to focus on the upper-middle class and marketed their tudor-style houses at between $9,100 and $18,500. Private commissions and smaller subdivisions carried the firm through the remainder of the prewar period.

In 1941 Levitt and Sons received a government contract for 1,600 (later increased to 2,350) war worker's homes in Norfolk, Virginia. The effort was a nightmare, but the brothers learned how to lay dozens of concrete foundations in a single day and to preassemble uniform walls and roofs. Additional contracts for more federal housing in Portsmouth, Virginia, and for barracks for shipyard workers at Pearl Harbor provided supplemental experience, as did William's service with the Navy Seabees from 1943 to 1945. Thus, the Levitts were among the nation's largest home builders even before construction of the first Levittown.

Returning to Long Island after the war, the Levitts built 2,250 houses in Roslyn in 1946 in the $17,500 to $23,500 price range, well beyond the means of the average veteran. In that same year, however, they began the acquisition of 4,000 acres of potato farms in the Town of Hempstead, where they planned the biggest private housing project in American history.

The formula for Island Trees, soon renamed Levittown, was simple. After bulldozing the land and removing the trees, trucks carefully dropped off building materials at precise 60-foot intervals. Each house was built on a concrete slab (no cellar); the floors were of asphalt and the walls of composition rock-board. Plywood replaced ¾-inch strip lap, ¾-inch double lap was changed to ⅜-inch for roofing, and the horse and scoop were replaced by the bulldozer. New power hand tools like saws, routers, and nailers helped increase worker productivity. Freight cars loaded with lumber went directly into a cutting yard where one man cut parts for ten houses in one day.

The construction process itself was divided into twenty-seven distinct steps—beginning with laying the foundation and ending with a clean sweep of the new home. Crews were trained to do one job—one day the white-paint men, then the red-paint men, then the tile layers. Every possible part, and especially the most difficult ones, were preassembled in central shops, whereas most builders did it on site. Thus, the Levitts reduced the skilled component to 20–40 percent. The five-day work week was standard, but they were the five days during which building was possible; Saturday and Sunday were considered to be the days when it rained. In the process, the Levitts defied unions and union work rules (against spray painting, for example) and insisted that subcontractors work only for them. Vertical integration also meant that the firm made its own concrete, grew its own timber, and cut its own lumber. It also bought all appliances from wholly owned subsidiaries. More than thirty houses went up each day at the peak of production.

Initially limited to veterans, this first "Levittown" was twenty-five miles east of Manhattan and particularly attractive to new families that had been formed during and just after the war. Squashed in with their in-laws or in tiny apartments where landlords frowned on children, the GI's looked upon Levittown as the answer to their most pressing need. Months before the first three hundred Levitt houses were occupied in October 1947, customers stood in line for the four-room Cape Cod box renting at sixty dollars per month. The first eighteen hundred houses were initially available only for rental, with an option to buy after a year's residence. Because the total for mortgage, interest, principal, and taxes was *less* than the rent, almost everyone bought; after 1949 all units were for sale only. So many of the purchasers were young families that the first issue of *Island Trees*, the community newspaper, opined that "our lives are held closely together because most of us are within the same age bracket, in similar income groups, live in almost identical houses and have common problems," and so many babies were born to them that the suburb came to be known as "Fertility Valley" and "The Rabbit Hutch."

Ultimately encompassing more than 17,400 separate houses and 82,000 residents, Levittown was the largest housing development ever put up by a single builder, and it served the American dream-house market at close to the lowest prices the industry could attain. The typical Cape Cod was down-to-earth and unpretentious; the intention was not to stir the imagination, but to provide the best shelter at the least price. Each dwelling included a twelve-by-sixteen-foot living-room with a fireplace, one bath, and two bedrooms (about 750 square feet), with easy expansion possibilities upstairs in the unfinished attic or outward into the yard. Most importantly, the floor plan was practical and well-designed, with the kitchen moved to the front of the house near the entrance so that mothers could watch their children from kitchen windows and do their washing and cooking with a minimum of movement. Similarly, the living room was placed in the rear and given a picture window overlooking the back yard. This early Levitt house was as basic to post World War II suburban development as the Model T had been to the automobile. In each case, the actual design features were less important than the fact that they were mass-produced and thus priced within the reach of the middle class.

William Jaird Levitt, who assumed primary operating responsibility for the firm soon after the war, disposed of houses as quickly as other men disposed of cars. Pricing his Cape Cods at $7,990 (the earliest models went for $6,990) and his ranches at $9,500, he promised no down payment, no closing costs, and "no hidden extras." With FHA and VA "production advances," Levitt boasted the largest line of credit ever offered a private home builder. He simplified the paperwork required for purchase and reduced the entire financing and titling transaction to two half-hour steps. His full-page advertisements offered a sweetener to eliminate lingering resistance—a Bendix washer was included in the purchase price. Other inducements included an eight-inch television set (for which the family would pay for the next thirty years). So efficient was the operation that *Harper's Magazine* reported in 1948 that Levitt undersold his nearest competition by $1,500 and still made a $1,000 profit on each house. As *New York Times'* architecture critic Paul Goldberger has noted, "Levittown houses were social creations more than architectural ones—they turned the detached, single-family house from a distant dream to a real possibility for thousands of middle-class American families."

Buyers received more than shelter for their money. When the initial families arrived with their baby strollers and play pens, there were no trees, schools, churches, or private telephones. Grocery shopping was a planned adventure, and picking up the mail required sloshing through the mud to Hicksville. The Levitts planted apple, cherry, and evergreen trees on each plot, however, and the development ultimately assumed a more parklike appearance. To facilitate development as a garden community, streets were curvilinear (and invariably called "roads" or "lanes"), and through traffic was shunted to peripheral thoroughfares. Nine swimming pools, sixty playgrounds, ten baseball diamonds, and seven "village greens" provided open space and recreational opportunities. The Levitts forbade fences (a practice later ignored) and permitted outdoor clothes drying only on specially designed, collapsible racks. They even supervised lawn-cutting for the first few years—doing the jobs themselves if necessary and sending the laggard families the bill.

Architectural critics, many of whom were unaccustomed to the tastes or resources of moderate-income people, were generally unimpressed by the repetitious houses on 60-by-100-foot "cookie cutter lots" and referred to Levittown as "degraded in conception and impoverished in form." From the Wantagh Parkway, the town stretched away to the east as far as the eye could see, house after identical house, a horizon broken only by telephone poles. Paul Goldberger, who admired the individual designs, thought that the whole was "an urban planning disaster," while Lewis Mumford complained that Levittown's narrow range of house type and income range resulted in a one-class community and a backward design. He noted that the Levitts used "new-fashioned methods to compound old-fashioned mistakes."

But Levittown was a huge popular success where it counted—in the marketplace. On a single day in March 1949, fourteen hundred contracts were drawn, some with families that had been in line for four days. "I truly loved it," recalled one early resident. "When they built the Village Green, our big event was walking down there for ice cream."

In the 1950s the Levitts shifted their attention from Long Island to an equally large project near Philadelphia. Located on former broccoli and spinach farms in lower Bucks County, Pennsylvania, this new Levittown was built within a few miles of the new Fairless Works of the United States Steel Corporation, where the largest percentage of the community's residents were employed. It was composed on eight master blocks, each of about one square mile and focusing on its own recreational facilities. Totaling about 16,000 homes when completed late in the decade, the town included light industry and a big, 55-acre shopping center. According to Levitt, "We planned every foot of it—every store, filling station, school, house, apartment, church, color, tree, and shrub."

In the 1960s, the Levitt forces shifted once again, this time to Willingboro, New Jersey, where a third Levittown was constructed within distant commuting range of Philadelphia. This last town was the focus of Herbert Gans's well-known account of *The Levittowners*. The Cape Cod remained the basic style, but Levitt improved the older models to resemble more closely the pseudo-colonial design that was so popular in the Northeast.

If imitation is the sincerest form of flattery, then William Levitt has been much honored in the past forty years. His replacement of basement foundations with the radiantly heated concrete slab was being widely copied as early as 1950. Levitt did not actually pioneer many of the mass-production techniques—the use of plywood, particle board, and gypsum board, as well as power hand tools like saws, routers, and nailers, for example—but his developments were so widely publicized that in every large metropolitan area, large builders appeared who adopted similar methods—Joseph Kelly in Boston, Frank White in Portland, Louis H. Boyar and Fritz B. Burns in Los Angeles, Del Webb in Phoenix, William G. Farrington in Houston, Franklin L. Burns in Denver, Wallace E. Johnson in Memphis, Ray Ellison in San Antonio, Maurice Fishman in Cleveland, Waverly Taylor in Washington, Irving Blietz and Phillip Klutznick in Chicago, John Mowbray in Baltimore, and Carl Gellert and Ellie Stoneman in San Francisco, to name just the more well-known builders.

FHA and VA programs made possible the financing of their immense developments. Title VI of the National Housing Act of 1934 allowed a builder to insure 90 percent of the mortgage of a house costing up to nine thousand dollars. Most importantly, an ambitious entrepreneur could get an FHA "commitment" to insure the mortgage, and then use the "commitment" to sign himself up as a temporary mortgagor. The mortgage lender (a bank or savings and loan institution) would then make "production advances" to the contractor as the work progressed, so that the builder needed to invest very little of his own hard cash. Previously, even the largest builders could not bring together the capital to undertake thousand-house developments. FHA alone insured three thousand houses in Henry J. Kaiser's Panorama City, California; five thousand in Frank Sharp's Oak Forest; and eight thousand in Klutznick's Park Forest project.

Characteristics of Postwar Suburbs

However financed and by whomever built, the new subdivisions that were typical of American urban development between 1945 and 1973 tended to share five common characteristics. The first was peripheral location. A Bureau of Labor Statistics survey of home building in 1946–1947 in six metropolitan regions determined that the suburbs accounted for at least 62 percent of construction. By 1950 the national suburban growth rate was ten times that of central cities, and in 1954 the editors of *Fortune* estimated that 9 million people had moved to the suburbs in the previous decade. The inner cities did have some empty lots—serviced by sewers, electrical connections, gas lines, and streets—available for development. But the filling-in process was not amenable to mass production techniques, and it satisfied neither the economic nor the psychological temper of the times.

The few new neighborhoods that were located within the boundaries of major cities tended also to be on the open land at the edges of the built-up sections. In New York City, the only area in the 1946–1947 study where city construction was greater than that of the suburbs, the big growth was on the outer edges of Queens, a borough that had been largely undeveloped in 1945. In Memphis new development moved east out Summer, Poplar, Walnut Grove, and Park Avenues, where FHA and VA subdivisions advertised "No Down Payment" or "One Dollar Down" on giant billboards. In Los Angeles the fastest-growing American city in the immediate postwar period, the area of rapid building focused on the San Fernando Valley, a vast space that had remained largely vacant since its annexation to the city in 1915. In Philadelphia thousands of new houses were put up in farming areas that had legally been part of the city since 1854, but which in fact had functioned as agricultural settlements for generations.

The second major characteristic of the postwar suburbs was their relatively low density. In all except the most isolated instances, the row house completely lost favor; between 1946 and 1956, about 97 percent of all new single-family dwellings were completely detached, surrounded on every side by their own plots. Typical lot sizes were relatively uniform around the country, averaging between ⅕ (80 by 100 feet) and ⅒ (40 by 100 feet) of an acre and varying more with distance from the center than by region. Moreover, the new subdivisions allotted a higher proportion of their land area to streets and open spaces. Levittown, Long Island, for example, was settled at a density of 10,500 per square mile, which was about average for postwar suburbs but less than half as dense as the streetcar suburbs of a half-century earlier. This design of new neighborhoods on the assumption that residents would have automobiles meant that those without cars faced severe handicaps in access to jobs and shopping facilities.

This low-density pattern was in marked contrast with Europe. In war-ravaged countries east of the Rhine River, the concentration upon apartment buildings can be explained by the overriding necessity to provide shelter quickly for masses of displaced and homeless people. But in comparatively unscathed France, Denmark, and Spain, the single-family house was also a rarity. In Sweden, Stockholm committed itself to a suburban pattern along subway lines, a decision that implied a high-density residential pattern. Nowhere in Europe was there the land, the money, or the tradition for single-family home construction.

The third major characteristic of the postwar suburbs was their architectural similarity. A few custom homes were built for the rich, and mobile homes gained popularity with the poor and the transient, but for most American families in search of a new place to live some form of tract house was the most likely option. In order to simplify their production methods and reduce design fees, most of the larger developers offered no more than a half-dozen basic house plans, and some offered half that number. The result was a monotony and repetition that was especially stark in the early years of the subdivision, before the individual owners had transformed their homes and yards according to personal taste.

But the architectural similarity extended beyond the particular tract to the nation as a whole. Historically, each region of the country had developed an indigenous residential style—the colonial-style homes of New England, the row houses of Atlantic coastal cities, the famous Charleston town houses with their ends to the street, the raised plantation homes of the damp bayou country of Louisiana, and the encircled patios and massive walls of the Southwest. This regionalism of design extended to relatively small areas; early in the twentieth century a house on the South Carolina coast looked quite different from a house in the Piedmont a few hundred miles away.

This tradition began eroding after World War I, when the American dream house became, as already noted, the Cape Cod cottage, a quaint one-and-a-half-story dwelling. This design remained popular into the post-World War II years, when Levittown featured it as a bargain for veterans. In subsequent years, one fad after another became the rage. First, it was the split-level, then the ranch, then the modified colonial. In each case, the style tended to find support throughout the continent, so that by the 1960s the casual suburban visitor would have a difficult time deciphering whether she was in the environs of Boston or Dallas.

The ranch style, in particular, was evocative of the expansive mood of the post-World War II suburbs and of the disappearing regionality of style. It was almost as popular in Westchester County as in Los Angeles County. Remotely derived from the adobe dwellings of the Spanish colonial tradition and more directly derived from the famed prairie houses of Frank Lloyd Wright, with their low-pitched roofs, deep eaves, and pronounced horizontal lines, the typical ranch style houses of the 1950s were no larger than the average home a generation earlier. But the one-level ranch house suggested spacious living and an easy relationship with the outdoors. Mothers with small children did not have to contend with stairs. Most importantly, the postwar ranch home represented newness. In 1945 the publisher of the *Saturday Evening Post* reported that only 14 percent of the population wanted to live in an apartment or a "used" house. Whatever the style, the post-World War II house, in contrast to its turn-of-the-century predecessor, had no hall, no parlor, no stairs, and no porch. And the portion of the structure that projected farthest toward the street was the garage.

The fourth characteristic of post-World War II housing was its easy availability and thus its reduced suggestion of wealth. To be sure, upper-income suburbs and developments sprouted across the land, and some set high standards of style and design. Typically, they offered expansive lots, spacious and individualized designs, and affluent neighbors. But the most important income development of the period was the lowering of the threshhold of purchase. At every previous time in American history, and indeed for the 1980s as well, the successful acquisition of a family home required savings and effort of a major order. After World War II, however, because of mass-production techniques, government financing, high wages, and low interest rates, it was quite simply cheaper to buy new housing in the suburbs than it was to reinvest in central city properties or to rent at the market price.

The fifth and perhaps most important characteristic of the postwar suburb was economic and racial homogeneity. The sorting out of families by income and color began even before the Civil War and was stimulated by the growth of the factory system. This pattern was noticeable in both the exclusive Main Line suburbs of Philadelphia and New York and in the more bourgeois streetcar developments which were part of every city. The automobile accentuated this discriminatory "Jim Crow" pattern. In Atlanta where large numbers of whites flocked to the fast-growing and wealthy suburbs north of the city in the 1920s, Howard L. Preston has reported that: "By 1930, if racism could be measured in miles and minutes, blacks and whites were more segregated in the city of Atlanta than ever before." But many pre-1930 suburbs—places like Greenwich, Connecticut; Englewood, New Jersey; Evanston, Illinois; and Chestnut Hill, Massachusetts—maintained an exclusive image despite the presence of low-income or minority groups living in slums near or within the community.

The post-1945 developments took place against a background of the decline of factory-dominated cities. What was unusual in the new circumstances was not the presence of discrimination—Jews and Catholics as well as blacks had been excluded from certain neighborhoods for generations—but the thoroughness of the physical separation which it entailed. The Levitt organization, which was no more culpable in this regard than any other urban or suburban firm, publicly and officially refused to sell to blacks for two decades after the war. Nor did resellers deal with minorities. As William Levitt explained, "We can solve a housing problem, or we can try to solve a racial problem. But we cannot combine the two." Not surprisingly, in 1960 not a single one of the Long Island Levittown's 82,000 residents was black.

The economic and age homogeneity of large subdivisions and sometimes entire suburbs was almost as complete as the racial distinction. Although this tendency had been present even in the nineteenth century, the introduction of zoning—beginning with a New York City ordinance in 1916—served the general purpose of preserving residential class segregation and property values. In theory zoning was designed to protect the interests of all citizens by limiting land speculation and congestion. And it was popular. Although it represented an extraordinary growth of municipal power, nearly everyone supported zoning. By 1926 seventy-six cities had adopted ordinances similar to that of New York. By 1936, 1,322 cities (85 percent of the total) had them, and zoning laws were affecting more property than all national laws relating to business.

In actuality zoning was a device to keep poor people and obnoxious industries out of affluent areas. And in time, it also became a cudgel used by suburban areas to whack the central city. Advocates of land-use restrictions of overwhelming proportion were residents of the fringe. They sought through minimum lot and set-back requirements to insure that only members of acceptable social classes could settle in their privileged sanctuaries. Southern cities even used zoning to enforce racial segregation. And in suburbs everywhere, North and South, zoning was used by the people who already lived within the arbitrary boundaries of a community as a method of keeping everyone else out. Apartments, factories, and "blight," euphemisms for blacks and people of limited means, were rigidly excluded.

While zoning provided a way for suburban areas to become secure enclaves for the well-to-do, it forced the city to provide economic facilities for the whole area and homes for people the suburbs refused to admit. Simply put, land-use restrictions tended to protect residential interests in the suburbs and commercial interests in the cities because the residents of the core usually lived on land owned by absentee landlords who were more interested in financial returns than neighborhood preferences. For the man who owned land but did not live on it, the ideal situation was to have his parcel of earth zoned for commercial or industrial use. With more options, the property often gained in value. In Chicago, for example, three times as much land was zoned for commercial use as could ever have been profitably employed for such purposes. This overzoning prevented inner-city residents from receiving the same protection from commercial incursions as was afforded suburbanites. Instead of becoming a useful tool for the rational ordering of land in metropolitan areas, zoning became a way for suburbs to pirate from the city only its desirable functions and residents. Suburban governments became like so many residential hotels, fighting for the upper-income trade while trying to force the deadbeats to go elsewhere.

Because zoning restrictions typically excluded all apartments and houses and lots of less than a certain number of square feet, new home purchasers were often from a similar income and social group. In this regard, the postwar suburbs were no different from many nineteenth-century neighborhoods when they were first built. Moreover, Levittown was originally a mix of young professionals and lower-middle-class blue-collar workers.

As the aspiring professionals moved out, however, Levittown became a community of the most class-stratifying sort possible. This phenomenon was the subject of one of the most important books of the 1950s. Focusing on a 2,400-acre project put up by the former Public Housing Administrator Phillip Klutznick, William H. Whyte's *The Organization Man* sent shudders through armchair sociologists. Although Whyte found that Park Forest, Illinois, offered its residents "leadership training" and an "ability to chew on real problems," the basic portrait was unflattering. Reporting excessive conformity and mindless conservatism, he showed Park Foresters to be almost interchangeable as they fought their way up the corporate ladder, and his "organization man" stereotype unfortunately became the norm for judging similar communities throughout the nation.

By 1961, when President John F. Kennedy proclaimed his New Frontier and challenged Americans to send a man to the moon within the decade, his countrymen had already remade the nation's metropolitan areas in the short space of sixteen years. From Boston to Los Angeles, vast new subdivisions and virtually new towns sprawled where a generation earlier nature had held sway. In an era of low inflation, plentiful energy, federal subsidies, and expansive optimism, Americans showed the way to a more abundant and more perfect lifestyle. Almost every contractor-built, post-World War II home had central heating, indoor plumbing, telephones, automatic stoves, refrigerators, and washing machines.

There was a darker side to the outward movement. By making it possible for young couples to have separate households of their own, abundance further weakened the extended family in America and ordained that most children would grow up in intimate contact only with their parents and siblings. The housing arrangements of the new prosperity were evident as early as 1950. In that year there were 45,983,000 dwelling units to accommodate the 38,310,000, families in the United States and 84 percent of American households reported less than one person per room.

Critics regarded the peripheral environment as devastating particularly to women and children. The suburban world was a female world, especially during the day. Betty Friedan's 1968 classic *The Feminine Mystique* challenged the notion that the American dream home was emotionally fulfilling for women. As Gwendolyn Wright has observed, their isolation from work opportunities and from contact with employed adults led to stifled frustration and deep psychological problems. Similarly, Sidonie M. Gruenberg warned in the *New York Times Magazine* that "Mass produced, standardized housing breeds standardized individuals, too—especially among youngsters." Offering neither the urbanity and sophistication of the city nor the tranquility and repose of the farm, the suburb came to be regarded less as an intelligent compromise than a cultural, economic, and emotional wasteland. No observer was more critical than Lewis Mumford, however. In his 1961 analysis of *The City in History*, which covered the entire sweep of civilization, the famed author reiterated sentiments he had first expressed more than four decades earlier and scorned the new developments which were surrounding every American city:

> In the mass movement into suburban areas a new kind of community was produced, which caricatured both the historic city and the archetypal suburban refuge: a multitude of uniform, unidentifiable houses, lined up inflexibly, at uniform distances, on uniform roads, in a treeless communal waste, inhabited by people of the same class, the same income, the same age group, witnessing the same television performances, eating the same tasteless pre-fabricated foods, from the same freezers, conforming in every outward and inward respect to a common mold, manufactured in the central metropolis. Thus, the ultimate effect of the suburban escape in our own time is, ironically, a low-grade uniform environment from which escape is impossible.

Secondly, because the federally supported home-building boom was of such enormous proportions, the new houses of the suburbs were a major cause of the decline of central cities. Because FHA and VA terms for new construction were so favorable as to make the suburbs accessible to almost all white, middle-income families, the inner-city housing market was deprived of the purchasers who could perhaps have supplied an appropriate demand for the evacuated neighborhoods.

The young families who joyously moved into the new homes of the suburbs were not terribly concerned about the problems of the inner-city housing market or the snobbish views of Lewis Mumford and other social critics. They were concerned about their hopes and their dreams. They were looking for good schools, private space, and personal safety, and places like Levittown could provide those amenities on a scale and at a price that crowded city neighborhoods, both in the Old World and in the new, could not match. The single-family tract house—post-World War II style—whatever its aesthetic failings, offered growing families a private haven in a heartless world. If the dream did not include minorities or the elderly, if it was accompanied by the isolation of nuclear families, by the decline of public transportation, and by the deterioration of urban neighborhoods, the creation of good, inexpensive suburban housing on an unprecedented scale was a unique achievement in the world.

CHAPTER 10

The Making of the Atomic Bomb

Richard Rhodes

On the morning of August 6, 1945, over the Japanese city of Hiroshima, the U.S. Air Force bomber *Enola Gay* dropped the first combat atomic bomb. Upon learning of its successful detonation, President Harry Truman declared, "This is the greatest thing in history."

Three days later, a second atomic weapon was dropped upon Nagasaki. The carnage in both cities was horrifying—140,000 dead or soon to die in Hiroshima, another 70,000 in Nagasaki.

The bombs brought an almost instantaneous end to the war. On August 15, Japan's Emperor Hirohito addressed his subjects over radio: "The enemy has begun to employ a new and most cruel bomb, the power of which to do damage is indeed incalculable, taking the toll of many innocent lives. . . . This is the reason why We have ordered [acceptance of the Allied terms of surrender]."

Taking as his subject the history of the development and deployment of these first atomic weapons, Richard Rhodes explores the events of the preceding years that led to the beginning of the atomic age. The silent question that underlies his work and that of hundreds of others who have also reflected on these events is, Could history have taken an alternative course?

In this selection, we examine the thoughts and actions of the Truman administration and its principal advisers in an important time period, the spring of 1945, when Truman took over the presidency upon Roosevelt's death. The war in the European theater ended soon, in early May, and the new president's attention was focused on the Pacific—and on our ostensible allies, the Russians.

Rhodes gives us detailed portraits of several of the major actors here: Truman's respected secretary of war, Henry Stimson, and the wily "assistant president," James "Jimmy" Byrnes. Truman's own temperament is also revealed in some detail, too. These details bear historical significance for helping us to explain how these men—all laypersons— groped in different ways to adapt the technical miracle of the atomic bomb to the nontechnical world of foreign relations and human affairs.

The bomb's development was the task of the Manhattan Project, the $2 billion program that was shrouded in closely guarded secrecy and headed by General Leslie Groves. Even as vice president, Truman had been kept in the dark about the atomic bomb, and had to be briefed immediately after he was sworn in as president.

During that spring of 1945, Truman was asked to respond to several challenges posed by the imminent successful completion of the bomb. One concern raised early by some of the most eminent scientists working on the Manhattan Project was the bomb's challenge to postwar relations between the United States, the Soviet Union, and the rest of the world. Niels Bohr had urged that the Soviet Union be told, before the first bombs were built, that a bomb project was under way. He believed that early disclosure would demonstrate good faith and could lead to negotiations on postwar arms control, heading off a nuclear arms race.

Fellow scientists Vannevar Bush and James Bryant Conant had refined Bohr's plea into a detailed proposal for the Roosevelt administration. They explicitly recommended that the United States sacrifice some portion of its national sovereignty in exchange for effective international control. Henry Stimson had remained unpersuaded, however, and no action concerning international sharing and controls had been taken while Roosevelt was alive.

When Truman took office, a new opportunity seemed to present itself to revive the proposals of Bush and Conant, but Rhodes argues that the opportunity was lost because Truman almost immediately took a confrontational stance with the Russians on the other issues. Henry Stimson, who briefed the president about the proposals to share information with other countries, was himself unenthusiastic.

The scene shifts, and the author shows a second and more short-term challenge posed by the bomb's nearing completion: the question of deciding where on Japan it should be dropped. The Target Committee carried out its deliberations with matter-of-fact dispassion. The scene shifts again, and we learn of a third challenge that the bomb presented, that of assembling the proper advisers for what would be called the Interim Committee, which was charged with the responsibility of advising Truman about the decision to use the bomb. This committee would later take up such questions as whether the first atomic bomb should be detonated on an uninhabited island with representatives of the Japanese government present nearby to witness the weapon's destructive power. Elsewhere in his book, Rhodes tells of how the Interim Committee would decide in the end to recommend to the president that the bomb be used against Japan as soon as possible and without prior warning.

The Target Committee carried forward its work, and the list of target candidates was narrowed for "Little Boy" and "Fat Man," the nicknames given to the bombs that were being readied. The Interim Committee took up Niels Bohr's proposal that all countries freely exchange scientific information and permit inspection of all laboratories—and military installations. The debate about the propriety of sharing scientific data with the Soviets appears to have pivoted not on high-minded principles of international cooperation, but rather on differing estimations of how long it would take Soviet scientists to catch up on their own.

The United States treated its two major allies differently. The Interim Committee remained unconvinced of the need to share our nuclear data with the Soviets, but the United States had already made its other ally, Great Britain, a nuclear partner under the terms of the Quebec Agreement. Secretary of State Byrnes was not subject, however, to the distorting influence of Anglophilia, and he was unhappy to learn that the United States had given away so much to Britain for so little in return. Whether dealing with the Soviets or with the British, Byrnes was consistent in his insistence on our always demanding something in exchange, or quid pro quo.

In August, the bombs were used without the prior warning that some Manhattan Project scientists had urged, and without the United States offering unilaterally to share the Anglo-American nuclear monopoly with the Soviet Union. Some historians have attributed the origins of the Cold War to American decisions made in these months of the spring and summer of 1945. They argue that Japan had already been effectively devastated without the dropping of the atomic bombs and would have surrendered in short order anyhow; consequently, the planned amphibious American invasion of the Japanese home islands would never have needed to come to pass. In this view, the United States was practicing what has been dubbed "atomic diplomacy": The intended primary audience for the bombs was not the Japanese but the Russians. Other historians have countered by pointing to many indications that the Japanese were not so close to surrendering.

These awful questions about the use of nuclear weapons of course hang in the air today, when nuclear arsenals on both sides bristle with thousands of warheads. In an eerie sense, we face today a situation that resembles that of early 1945—here in the United States, we once again seem to have a momentary technological edge over the Soviet Union and possess secret information concerning an entirely new form of lethal weapon, the space-based laser weapons proposed in the Strategic Defense Initiative, "Star Wars." As the country debates SDI, it confronts old questions that go beyond simply the trustworthiness of the Soviets, or the technical problems that may make development of the actual weapons prohibitively expensive. It confronts the deeper difficulty of resisting technological temptation—If something can be done, can we choose *not* to do it?

LIFE AND DEATH

Within twenty-four hours of Franklin Roosevelt's death two men told Harry Truman about the atomic bomb. The first was Henry Lewis Stimson, the upright, white-haired, distinguished Secretary of War. He spoke to the newly sworn President following the brief cabinet meeting Truman called after taking the oath of office on the evening of the day Roosevelt died. "Stimson told me," Truman reports in his memoirs, "that he wanted me to know about an immense project that was underway—a project looking to the development of a new explosive of almost unbelievable destructive power. That was all he felt free to say at the time, and his statement left me puzzled. It was the first bit of information that had come to me about the atomic bomb, but he gave me no details."

Truman had known of the Manhattan Project's existence since his wartime Senate work as chairman of the Committee to Investigate the National Defense Program, when he had attempted to explore the expensive secret project's purpose and had been rebuffed by the Secretary of War himself. That a senator of watchdog responsibility and bulldog tenacity would call off an investigation into unaccounted millions of dollars in defense-plant construction on Stimson's word alone gives some measure of the quality of the Secretary's reputation.

Stimson was seventy-seven years old when Truman assumed the Presidency. He could remember stories his great-grandmother told him of her childhood talks with George Washington. He had attended Phillips Andover when the tuition at the distinguished New England preparatory school was sixty dollars a year and students cut their own firewood. He had graduated from Yale College and Harvard Law School, had served as Secretary of War under William Howard Taft, as Governor General of the Philippines under Calvin Coolidge, as Secretary of State under Herbert Hoover. Roosevelt had called him back to active service in 1940 and with able assistance especially from George Marshall and despite insomnia and migraines that frequently laid him low he had built and administered the most powerful military organization in the history of the world. He was a man of duty and of rectitude. "The chief

SOURCE: From *The Making of the Atomic Bomb* by Richard Rhodes. Copyright © 1986 by Richard Rhodes. Reprinted by permission of Simon & Schuster.

lesson I have learned in a long life," he wrote at the end of his career, "is that the only way you can make a man trustworthy is to trust him; and the surest way to make him untrustworthy is to distrust him and show your distrust." Stimson sought to apply the lesson impartially to men and to nations. In the spring of 1945 he was greatly worried about the use and consequences of the atomic bomb.

The other man who spoke to Truman, on the following day, April 13, was James Francis Byrnes, known as Jimmy, sixty-six years old, a private citizen of South Carolina since the beginning of April but before then for three years what Franklin Roosevelt had styled "assistant President": Director of Economic Stabilization and then Director of War Mobilization, with offices in the White House. While FDR ran the war and foreign affairs, that is, Byrnes had run the country. "Jimmy Byrnes . . . came to see me," writes Truman of his second briefing on the atomic bomb, "and even he told me few details, though with great solemnity he said that we were perfecting an explosive great enough to destroy the whole world." Then or soon afterward, before Truman met with Stimson again, Byrnes added a significant twist to his tale: "that in his belief the bomb might well put us in a position to dictate our own terms at the end of the war."

At that first Friday meeting Truman asked Byrnes to transcribe his shorthand notes on the Yalta Conference, three months past, which Byrnes had attended as one of Roosevelt's advisers and about which Truman, merely the Vice President then, knew little. Yalta represented nearly all Byrnes' direct experience of foreign affairs. It was more than Truman had. Under the circumstances the new President found it sufficient and informed his colleague that he meant to make him Secretary of State. Byrnes did not object. He insisted that he be given a free hand, however, as Roosevelt had given him in domestic affairs, and Truman agreed.

"A small, wiry, neatly made man," a team of contemporary observers describes Jimmy Byrnes, "with an odd, sharply angular face from which his sharp eyes peer out with an expression of quizzical geniality." Dean Acheson, then an Assistant Secretary of State, thought Byrnes overconfident and insensitive, "a vigorous extrovert, accustomed to the lusty exchange of South Carolina politics." Truman assayed the South Carolinian most shrewdly a few months after their April discussion in a private diary he intermittently kept:

> Had a long talk with my able and conniving Secretary of State. My but he has a keen mind! And he is an honest man. But all country politicians are alike. They are sure all other politicians are circuitous in their dealings. When they are told the straight truth, unvarnished, it is never believed— an asset *sometimes*.

A politician's politician, Byrnes had managed in his thirty-two years of public life to serve with distinction in all three branches of the federal government. He was self-made from the ground up. His father died before he was born. His mother learned dressmaking to survive. Young Jimmy found work at fourteen, his last year of formal education, in a law office, but in lieu of classroom study one of the law partners kindly guided him through a comprehensive reading list. His mother in the meantime taught him shorthand and in 1900, at twenty-one, he earned appointment as a court reporter. He read for the law under the judge whose circuit he reported and passed the bar in 1904. He ran first, in 1908, for solicitor, the South Carolina equivalent of district attorney, and made himself known prosecuting murderers. More than forty-six stump debates won him election to Congress in 1910; in 1930, after fourteen years in the House and five years out of office, he was elected to the Senate. By then he was already actively promoting Franklin Roosevelt's approaching presidential bid. Byrnes served as one of the candidate's speechwriters during the 1932 campaign and afterward worked hard as Roosevelt's man in the Senate to push through the New Deal. His reward, in 1941, was a seat on the United States Supreme Court, which he resigned in 1942 to move to the White House to take over operating the complicated wartime emergency program of wage and price controls, the assistant Presidency of which Roosevelt spoke

In 1944 everyone understood that Roosevelt's fourth term would be his last. The man he selected for Vice President would therefore almost certainly take the Democratic Party presidential nomination in 1948. Byrnes expected to be that man and Roosevelt encouraged him. But the assistant President was a conservative Democrat from the Deep South, and at the last minute Roosevelt compromised instead on the man from Missouri, Harry S. Truman. "I freely admit that I was disappointed," Byrnes writes with understatement approaching lockjaw, "and felt hurt by President Roosevelt's action." He made a point of visiting the European front with George Marshall in September 1944, in the midst of the presidential campaign; when he returned FDR had to appeal to him formally by letter—a document Byrnes could show around—to endorse the ticket with a speech.

Byrnes undoubtedly regarded Truman as a usurper: if not Truman but he had been Roosevelt's choice he would be President of the United States now. Truman knew Byrnes' attitude but needed the old pro badly to help him run the country and face the world. Hence the prize of State. The Secretary of State was the highest-ranking member of the cabinet and under the rules of succession then obtaining was the officer next in line for the Presidency as well when the Vice Presidency was vacant. Short of the Presidency itself, State was the most powerful office Truman had to give.

Vannevar Bush and James Bryant Conant had needed months to convince Henry Stimson to take up consideration of the bomb's challenge in the postwar era. He had not been ready in late October 1944 when Bush pressed him for action and he had not been ready in early December when Bush pressed him again. By then Bush knew what he thought the problem needed, however:

> We proposed that the Secretary of War suggest to the President the establishment of a committee or commission with the duty of preparing plans. These would include the drafting of legislation and the drafting of appropriate releases to be made public at the proper time. . . . We were all in agreement that the State Department should now be brought in.

Stimson allowed one of his trusted aides, Harvey H. Bundy, a Boston lawyer, father of William P. and McGeorge, at least to begin formulating a membership roster and list of duties for such a committee. But he did not yet know even in broad outline what basic policy to recommend. Bohr's ideas, variously diluted, floated by that time in the Washington air. Bohr had sought to convince the American government that only early discussion with the Soviet Union of the mutual dangers of a nuclear arms race could forestall such an arms race once the bomb became known. (He would try again in April to see Roosevelt; Felix Frankfurter and Lord Halifax, the British ambassador, would be strolling in a Washington park discussing Bohr's best avenue of approach when the bells of the city's churches began tolling the news of the President's death.) Apparently no one within the executive branch was sufficiently convinced of the *inevitability* of Bohr's vision. Stimson was as wise as any man in the government, but late in December he cautioned Roosevelt that the Russians should earn the right to hear the baleful news:

> I told him of my views as to the future of S-1 [Stimson's code for the bomb] in connection with Russia: that I knew they were spying on our work but had not yet gotten any real knowledge of it and that, while I was troubled about the possible effect of keeping from them even now that work, I believed that it was essential not to take them into our confidence until we were sure to get a real quid pro quo from our frankness. I said I had no illusions as to the possibility of keeping permanently such a secret but that I did not think it was yet time to share it with Russia. He said he thought he agreed with me.

In mid-February, after talking again to Bush, Stimson confided to his diary what he wanted in exchange for news of the bomb. Bohr's conviction that only an open world modeled in some sense on the republic of science could answer the challenge of the bomb had drifted, in Bush's mind, to a proposal for an international pool of scientific research. Of such an arrangement Stimson wrote that "it would be inadvisable to put it into full force yet, until we had gotten all we could in Russia in the way of liberalization in exchange for S-1." That is, the quid pro quo Stimson thought the United States should demand from the Soviet Union was the democratization of its government. What for Bohr was the inevitable outcome of a solution to the problem of the bomb—an open world where differences in social and political conditions would be visible to everyone and therefore under pressure to improve—Stimson imagined should be a precondition to any initial change.

Finally in mid-March Stimson talked to Roosevelt, their last meeting. That talk came to no useful end. In April, with a new President in the White House, he prepared to repeat the performance.

In the meantime the men who had advised Franklin Roosevelt were working to convince Harry Truman of the increasing perfidy of the Soviet Union. Averell Harriman, the shrewd multimillionaire Ambassador to Moscow, had rushed to Washington to brief the new President. Truman says Harriman told him the visit was based on "the fear that you did not understand, as I had seen Roosevelt understand, that Stalin is breaking his agreements." To soften that condescension Harriman added that he feared Truman "could not have had time to catch up with all the recent cables." The self-educated Missourian prided himself on how many pages of documents he could chew through per day—he was a champion reader—and undercut Harriman's condescension breezily by instructing the ambassador to "keep on sending me long messages."

Harriman told Truman they were faced with a "barbarian invasion of Europe." The Soviet Union, he said, meant to take over its neighbors and install the Soviet system of secret police and state control. "He added that he was not pessimistic," the President writes, "for he felt that it was possible for us to arrive at a workable basis with the Russians. He believed that this would require a reconsideration of our policy and the abandonment of any illusion that the Soviet government was likely soon to act in accordance with the principles to which the rest of the world held in international affairs."

Truman was concerned to convince Roosevelt's advisers that he meant to be decisive. "I ended the meeting by saying, 'I intend to be firm in my dealings with the Soviet government.'" Delegates were arriving in San Francisco that April, for example, to formulate a charter for a new United Nations to replace the old and defunct League. Harriman asked Truman if he would "go ahead with the world organization." Three days later, having heard from Stalin in the meantime and met the arriving Soviet Foreign Minister, Vyacheslav Molotov, he retreated from realism to bluster. "He felt that our agreements with the Soviet Union had so far been a one-way street," an eyewitness recalls, "and that he could not continue; it was now or never. He intended to go on with the plans for San Francisco and if the Russians did not wish to join us they could go to Hell."

Stimson argued for patience. "In the big military matters," Truman reports him saying, "the Soviet government had kept its word and the military authorities of the United States had come to count on it. In fact. . .they had often done better than they had promised." Although George Marshall seconded Stimson's argument and Truman could not have had two more reliable witnesses, it was not counsel the new and untried President wanted to hear. Marshall added a crucial justification that Truman took to heart:

> He said from the military point of view the situation in Europe was secure but that we hoped for Soviet participation in the war against Japan at a time when it would be useful to us. The Russians had it within their power to delay their entry into the Far Eastern war until we had done all the dirty work. He was inclined to agree with Mr. Stimson that the possibility of a break with Russia was very serious.

Truman could hardly tell the Russians to go to hell if he needed them to finish the Pacific war. Marshall's justification for patience meant Stalin had the President over a barrel. It was not an arrangement Harry Truman intended to perpetuate.

He let Molotov know. They had sparred diplomatically at their first meeting; now the President attacked. The issue was the composition of the postwar government in Poland. Molotov discussed various formulas, all favoring Soviet dominance. Truman demanded the free elections that he understood had been agreed upon at Yalta: "I replied sharply that an agreement had been reached on Poland and that there was only one thing to do, and that was for Marshal Stalin to carry out the agreement in accordance with his word." Molotov tried again. Truman replied sharply again, repeating his previous demand. Molotov hedged once more. Truman proceeded to lay him low: "I expressed once more the desire of the United States for friendship with Russia, but I wanted it clearly understood that this could be only on a basis of the mutual observation of agreements and not on the basis of a one-way street." Those are hardly fighting words; Molotov's reaction suggests that the President spoke more pungently at the time:

> "I have never been talked to like that in my life," Molotov said.
> I told him, "Carry out your agreements and you won't get talked to like that."

If Truman felt better for the exchange, it disturbed Stimson. The new President had acted without knowledge of the bomb and its potentially fateful consequences. It was time and past time for a full briefing.

Truman agreed to meet with Stimson at noon on Wednesday, April 25. The President was scheduled to address the opening session of the United Nations conference in San Francisco by radio that evening. One more conditioning incident intervened; on Tuesday he received a communication from Joseph Stalin, "one of the most revealing and disquieting messages to reach me during my first days in the White House." Molotov had reported Truman's tough talk to the Soviet Premier. Stalin replied in kind. Poland bordered on the Soviet Union, he wrote, not on Great Britain or the United States. "The question [of] Poland had the same meaning for the security of the Soviet Union as the question [of] Belgium and Greece for the security of Great Britain"—but "the Soviet Union was not consulted when those governments were being established there" following the Allied liberation. The "blood of the Soviet people abundantly shed on the fields of Poland in the name of the liberation of Poland" demanded a Polish government friendly to Russia. And finally:

> I am ready to fulfill your request and do everything possible to reach a harmonious solution. But you demand too much of me. In other words, you demand that I renounce the interests of security of the Soviet Union, but I cannot turn against my country.

With this blunt challenge on his mind Truman received his Secretary of War.

Stimson had brought Groves along for technical backup but left him waiting in an outer office while he discussed issues of general policy. He began dramatically, reading from a memorandum:

> Within four months we shall in all probability have completed the most terrible weapon ever known in human history, one bomb of which could destroy a whole city.

We had shared the development with the British, Stimson continued, but we controlled the factories that made the explosive material "and no other nation could reach this position for some years." It was certain that we would not enjoy a monopoly forever, and "probably the only nation which could enter into production within the next few years is Russia." The world "in its present state of moral advancement compared with its technical development," the Secretary of War continued quaintly, "would be eventually at the mercy of such a weapon. In other words, modern civilization might be completely destroyed."

Stimson emphasized what John Anderson had emphasized to Churchill the year before: that founding a "world peace organization" while the bomb was still a secret "would seem to be unrealistic":

> No system of control heretofore considered would be adequate to control this menace. Both inside any particular country and between the nations

of the world, the control of this weapon will undoubtedly be a matter of the greatest difficulty and would involve such thorough-going rights of inspection and internal controls as we have never heretofore contemplated.

That brought Stimson to the crucial point:

> Furthermore, in the light of our present position with reference to this weapon, the question of sharing it with other nations and, if so shared, upon what terms, becomes a primary question of our foreign relations.

Bohr had proposed to inform other nations of the common dangers of a nuclear arms race. At the hands of Stimson and his advisers that sensible proposal had drifted to the notion that the issue was sharing the weapon itself. As Commander in Chief, as a veteran of the First World War, as a man of common sense, Truman must have wondered what on earth his Secretary of War was talking about, especially when Stimson added that "a certain moral responsibility" followed from American leadership in nuclear technology which the nation could not shirk "without very serious responsibility for any disaster to civilization which it would further." Was the United States morally obligated to give away a devastating new weapon of war?

Now Stimson called in Groves. The general brought with him a report on the status of the Manhattan Project that he had presented to the Secretary of War two days earlier. Both Stimson and Groves insisted Truman read the document while they waited. The President was restive. He had a threatening note from Stalin to deal with. He had to prepare to open the United Nations conference even though Stimson had just informed him that allowing the conference to proceed in ignorance of the bomb was a sham. A scene of darkening comedy followed as the proud man who had challenged Averell Harriman to keep sending him long messages tried to avoid public instruction in the minutiae of a secret project he had fought doggedly as a senator to investigate. Groves misunderstood completely:

> Mr. Truman did not like to read long reports. This report was not long, considering the size of the project. It was about twenty-four pages and he would constantly interrupt his reading to say, "Why, I don't like to read papers." And Mr. Stimson and I would reply: "Well we can't tell you this in any more concise language. This is a big project." For example, we discussed our relations with the British in about four or five lines. It was that much condensed. We had to explain all the processes and we might just say what they were and that was about all.

After the reading of the lesson, Groves notes, "a great deal of emphasis was placed on foreign relations and particularly on the Russian situation"—Truman reverting to his immediate problems. He "made it very definite," Groves adds for the record, "that he was in entire agreement with the necessity for the project."

The final point in Stimson's memorandum was the proposal Bush and Conant had initiated to establish what Stimson called "a select committee . . . for recommending action to the Executive and legislative branches of our government." Truman approved.

In his memoirs the President describes his meeting with Stimson and Groves with tact and perhaps even a measure of private humor: "I listened with absorbed interest, for Stimson was a man of great wisdom and foresight. He went into considerable detail in describing the nature and the power of the projected weapon. . . . Byrnes had already told me that the weapon might be so powerful as to be capable of wiping out entire cities and killing people on an unprecedented scale." That was when Byrnes had crowed that the new bombs might allow the United States to dictate its own terms at the end of the War. "Stimson, on the other hand, seemed at least as much concerned with the role of the atomic bomb in the shaping of history as in its capacity to shorten this war. . . . I thanked him for his enlightening presentation of this awesome subject, and as I saw him to the door I felt how fortunate the country was to have so able and so wise a man in its service." High praise, but the President was not sufficiently impressed at the outset with Stimson and Harriman to invite either man to accompany him to the next conference of the Big Three. Both found it necessary, when the time came, to invite themselves. Jimmy Byrnes went at the President's invitation and sat at the President's right hand.

Discussion between Truman and his various advisers was one level of discourse in the spring of 1945 on the uses of the atomic bomb. Another was joined two days after Stimson and Groves briefed the President when a Target Committee under Groves' authority met for the first time in Lauris Norstad's conference room at the Pentagon. Brigadier General Thomas F. Farrell, who would represent the Manhattan Project as Groves' deputy to the Pacific Command, chaired the committee; besides Farrell it counted two other Air Force officers—a colonel and a major—and five scientists, including John von Neumann and British physicist William G. Penney. Groves opened the meeting with a variant of his usual speech to Manhattan Project working groups: how important their duty was, how secret it must be kept. He had already discussed targets with the Military Policy Committee and now informed his Target Committee that it should propose no more than four.

Farrell laid down the basics: B-29 range for such important missions no more than 1,500 miles; visual bombing essential so that these untried and valuable bombs could be aimed with certainty and their effects photographed; probable targets "urban or industrial Japanese areas" in July, August or September; each mission to be given one primary and two alternate targets with spotter planes sent ahead to confirm visibility.

Most of the first meeting was devoted to worrying about the Japanese weather. After lunch the committee brought in the Twentieth Air Force's top meteorologist, who told them that June was the worst weather month in Japan; "a little improvement is present in July; a little bit better weather is present in August; September weather is bad." January was the best month, but no one intended to wait that long. The meteorologist said he could forecast a good day for bombing operations only twenty-four hours ahead, but he could give two days' notice of bad weather. He suggested they station submarines near the target areas to radio back weather readings.

Later in the afternoon they began considering targets. Groves had extended Farrell's guidelines:

> I had set as the governing factor that the targets chosen should be places the bombing of which would most adversely affect the will of the Japanese people to continue the war. Beyond that, they should be military in nature, consisting either of important headquarters or troop concentrations, or centers of production of military equipment and supplies. To enable us to assess accurately the effects of the bomb, the targets should not have been previously damaged by air raids. It was also desirable that the first target be of such size that the damage would be confined within it, so that we could more definitely determine the power of the bomb.

But such pristine targets had already become scarce in Japan. If the first choice the Target Committee identified at its first meeting was hardly big enough to confine the potential damage, it was the best the enemy had left to offer:

> Hiroshima is the largest untouched target not on the 21st Bomber Command priority list. Consideration should be given to this city.

"Tokyo," the committee notes continue, "is a possibility but it is now practically all bombed and burned out and is practically rubble with only the palace grounds left standing. Consideration is only possible here."

The Target Committee did not yet fully understand the level of authority it commanded. With a few words to Groves it could exempt a Japanese city from Curtis LeMay's relentless firebombing, preserving it through spring mornings of cherry blossoms and summer nights of wild monsoons for a more historic fate. The committee thought it took second priority behind LeMay rather than first priority ahead, and in emphasizing these mistaken priorities the colonel who reviewed the Twentieth Air Force's bombing directive for the committee revealed what the United States' policy in Japan in all its deadly ambiguity had become:

> It should be remembered that in our selection of any target, the 20th Air Force is operating primarily to laying waste all the main Japanese cities, and that they do not propose to save some important primary target for us if it interferes with the operation of the war from their point of view. Their existing procedure has been to bomb the hell out of Tokyo, bomb the

aircraft, manufacturing and assembly plants, engine plants and in general paralyze the aircraft industry so as to eliminate opposition to the 20th Air Force operations. The 20th Air Force is systematically bombing out the following cities with the prime purpose in mind of not leaving one stone lying on another:

Tokyo, Yokohama, Nagoya, Osaka, Kyoto,
Kobe, Yawata & Nagasaki.

If the Japanese were prepared to eat stones, the Americans were prepared to supply them.

The colonel also advised that the Twentieth Air Force planned to increase its delivery of conventional bombs steadily until it was dropping 100,000 tons a month by the end of 1945.

The group decided to study seventeen targets including Tokyo Bay, Yokohama, Nagoya, Osaka, Kobe, Hiroshima, Kokura, Fukuoka, Nagasaki and Sasebo. Targets already destroyed would be culled from the list. The weather people would review weather reports. Penney would consider "the size of the bomb burst, the amount of damage expected, and the ultimate distance at which people would be killed." Von Neumann would be responsible for computations. Adjourning its initial meeting the Target Committee planned to meet again in mid-May in Robert Oppenheimer's office at Los Alamos.

A third level of discourse on the uses of the bomb revealed itself as Henry Stimson assembled the committee that Bush and Conant had proposed to him and he had proposed in turn to the President. On May 1, the day German radio announced the suicide of Adolf Hitler in the ruins of Berlin, George L. Harrison, a special Stimson consultant and the president of the New York Life Insurance Company, prepared for the Secretary of War an entirely civilian committee roster consisting of Stimson as chairman, Bush, Conant, MIT president Karl Compton, Assistant Secretary of State William L. Clayton, Undersecretary of the Navy Ralph A. Bard and a special representative of the President whom the President might choose. Stimson modified the list to include Harrison as his alternate and carried it to Truman for approval on May 2. Truman agreed and Stimson apparently assumed his interest in the project, but the President significantly did not even bother to name his own man to the list. Stimson wrote in his diary that night:

> The President accepted the present members of the committee and said that they would be sufficient even without a personal representative of himself. I said I should prefer to have such a representative and suggested that he should be a man (a) with whom the President had close personal relations and (b) who was able to keep his mouth shut.

Truman had not yet announced his intention to appoint Byrnes Secretary of State because the holdover Secretary, Edward R. Stettinius, Jr., was heading the United States delegation to the United Nations in

San Francisco and the President did not want to undercut his authority there. But word of the forthcoming appointment had diffused through Washington. Acting on it, Harrison suggested that Stimson propose Byrnes. On May 3 Stimson did, "and late in the day the President called me up himself and said that he had heard of my suggestion and it was fine. He had already called up Byrnes down in South Carolina and Byrnes had accepted." Bundy and Harrison, Stimson told his diary, "were tickled to death." They thought their committee had acquired a second powerful sponsor. In fact they had just welcomed a cowbird into their nest.

Stimson sent out invitations the next day. He proposed calling his new group the Interim Committee to avoid appearing to usurp congressional prerogatives: "when secrecy is no longer required," he explained to the prospective members, "Congress might wish to appoint a permanent Post War Commission." He set the first informal meeting of the Interim Committee for May 9.

The membership would assemble in the wake of momentous change. The war in Europe had finally ground to an end. Supreme Allied Commander Dwight D. Eisenhower celebrated the victory on national radio the evening of Tuesday, May 8, 1945, V-E Day:

> I have the rare privilege of speaking for a victorious army of almost five million fighting men. They, and the women who have so ably assisted them, constitute the Allied Expeditionary Force that has liberated western Europe. They have destroyed or captured enemy armies totalling more than their own strength, and swept triumphantly forward over the hundreds of miles separating Cherbourg from Lübeck, Leipzig and Munich. . . .
>
> These startling successes have not been bought without sorrow and suffering. In this Theater alone 80,000 Americans and comparable numbers among their Allies, have had their lives cut short that the rest of us might live in the sunlight of freedom. . . .
>
> But, at last, *this* part of the job is done. No more will there flow from this Theater to the United States those doleful lists of death and loss that have brought so much sorrow to American homes. The sounds of battle have faded from the European scene.

Eisenhower had watched Colonel General Alfred Jodl sign the act of military surrender in a schoolroom in Rheims—the temporary war room of the Supreme Headquarters Allied Expeditionary Force—in the early morning hours of May 7. Eisenhower's aides had attempted then to draft a suitably eloquent message to the Combined Chiefs reporting the official surrender. "I tried one myself," Eisenhower's chief of staff Walter Bedell Smith remembers, "and like all my associates, groped for resounding phrases as fitting accolades to the Great Crusade and indicative of our dedication to the great task just accomplished." The Supreme Commander listened quietly for a time, thanked everyone for trying and dictated his own unadorned report:

The mission of this Allied force was fulfilled at 0241, local time, May 7th 1945.

Better to be brief, better than resounding phrases. Twenty million Soviet soldiers and civilians died of privation or in battle in the Second World War. Eight million British and Europeans died or were killed and another five million Germans. The Nazis murdered six million Jews in ghettos and concentration camps. Manmade death had ended thirty-nine million human lives prematurely; for the second time in half a century Europe had become a charnel house.

There remained the brutal conflict Japan had begun in the Pacific and refused despite her increasing destruction to end by unconditional surrender.

Officially Byrnes was retired to South Carolina. In fact he was visiting Washington surreptitiously, absorbing detailed evening briefings by State Department division chiefs at his apartment at the Shoreham Hotel. On the afternoon of V-E Day he spent two hours closeted alone with Stimson. Then Harrison, Bundy and Groves joined them. "We all discussed the function of the proposed Interim Committee," Stimson records. "During the meeting it became very evident what a tremendous help Byrnes would be as a member of the committee."

The next morning the Interim Committee met for the first time in Stimson's office. The gathering was preliminary, to fill in Byrnes, State's Clayton and the Navy's Bard on the basic facts, but Stimson made a point of introducing the former assistant President as Truman's personal representative. The membership was thus put on notice that Byrnes enjoyed special status and that his words carried extra weight.

The committee recognized that the scientists working on the atomic bomb might have useful advice to offer and created a Scientific Panel adjunct. Bush and Conant put their heads together and recommended Arthur Compton, Ernest Lawrence, Robert Oppenheimer and Enrico Fermi for appointment.

Between the first and second meetings of the Interim Committee its *Doppelgänger*, the Target Committee, met again for two days, May 10 and 11, at Los Alamos. Added to the full committee as advisers were Oppenheimer, Parsons, Tolman and Norman Ramsey and for part of the deliberations Hans Bethe and Robert Brode. Oppenheimer took control by devising and presenting a thorough agenda:

A. Height of Detonation
B. Report on Weather and Operations
C. Gadget Jettisoning and Landing
D. Status of Targets
E. Psychological Factors in Target Selection
F. Use Against Military Objectives
G. Radiological Effects

H. Coordinated Air Operations
I. Rehearsals
J. Operating Requirements for Safety of Airplanes
K. Coordination with 21st [Bomber Command] Progam

Detonation height determined how large an area would be damaged by blast and depended crucially on yield. A bomb detonated too high would expend its energy blasting thin air; a bomb detonated too low would expend its energy excavating a crater. It was better to be low than high, the committee minutes explain: "The bomb can be detonated as much as 40% below the optimum with a reduction of 24% in area of damage whereas a detonation [only] 14% above the optimum will cause the same loss in area." The discussion demonstrates how uncertain Los Alamos still was of bomb yield. Bethe estimated a yield range for Little Boy of 5,000 to 15,000 tons TNT equivalent. Fat Man, the implosion bomb, was anybody's guess: 700, 2,000, 5,000 tons? "With the present information the fuse would be set at 2,000 tons equivalent but fusing for the other values should be available at the time of final delivery. . . . Trinity data will be used for this gadget."

The scientists reported and the committee agreed that in an emergency a B-29 in good condition could return to base with a bomb. "It should make a normal landing with the greatest possible care. . . . The chances of [a] crash initiating a high order [i.e., nuclear] explosion are . . . sufficiently small [as to be] a justifiable risk." Fat Man could even survive jettisoning into shallow water. Little Boy was less forgiving. Since the gun bomb contained more than two critical masses of U235, seawater leaking into its casing could moderate stray neutrons sufficiently to initiate a destructive slow-neutron chain reaction. The alternative, jettisoning Little Boy onto land, might loose the U235 bullet down the barrel into the target core and set off a nuclear explosion. For temperamental Little Boy, the minutes note, unluckily for the aircrew, "the best emergency procedure that has so far been proposed is . . . the removal of the gun powder from the gun and the execution of a crash landing."

Target selection had advanced. The committee had refined its qualifications to three: "important targets in a large urban area of more than three miles diameter" that were "capable of being damaged effectively by blast" and were "likely to be unattacked by next August." The Air Force had agreed to reserve five such targets for atomic bombing. These included:

1. *Kyoto*—This target is an urban industrial area with a population of 1,000,000. It is the former capital of Japan and many people and industries are now being moved there as other areas are being destroyed. From the psychological point of view there is an advantage

that Kyoto is an intellectual center for Japan and the people there are more apt to appreciate the significance of such a weapon as the gadget. . . .

2. *Hiroshima*—This is an important army depot and port of embarkation in the middle of an urban industrial area. It is a good radar target and it is such a size that a large part of the city could be extensively damaged. There are adjacent hills which are likely to produce a focusing effect which would considerably increase the blast damage. Due to rivers it is not a good incendiary target.

The other three targets proposed were Yokohama, Kokura Arsenal and Niigata. An unsung enthusiast on the committee suggested a spectacular sixth target for consideration, but wiser heads prevailed: "The possibility of bombing the Emperor's palace was discussed. It was agreed that we should not recommend it but that any action for this bombing should come from authorities on military policy."

So the Target Committee sitting in Oppenheimer's office at Los Alamos under the modified Lincoln quotation that Oppenheimer had posted on the wall—THIS WORLD CANNOT ENDURE HALF SLAVE AND HALF FREE—remanded four targets to further study: Kyoto, Hiroshima, Yokohama and Kokura Arsenal.

The committee and its Los Alamos consultants were not unmindful of the radiation effects of the atomic bomb—its most significant difference in effect from conventional high explosives—but worried more about radiation danger to American aircrews than to the Japanese. "Dr. Oppenheimer presented a memo he had prepared on the radiological effect of the gadget. . . . The basic recommendations of this memo are (1) for radiological reasons no aircraft should be closer than 2½ miles to the point of detonation (for blast reasons the distance should be greater) and (2) aircraft must avoid the cloud of radio-active materials."

Since the expected yields of the bombs under discussion made them something less than city-busters, the Target Committee considered following Little Boy and Fat Man with conventional incendiary raids. Radioactive clouds that might endanger LeMay's follow-up crews worried the targeters, though they thought an incendiary raid delayed one day after an atomic bombing might be safe and "quite effective."

With a better sense for having visited Los Alamos of the weapons it was targeting, the Target Committee scheduled its next meeting for May 28 at the Pentagon.

Vannevar Bush thought the second Interim Committee meeting on May 14 produced "very frank discussions." The group, he decided, was "an excellent one." These judgments he passed along to Conant, who had been unable to attend. Stimson won approval of the Scientific Panel as constituted and discussed the possibility of assembling a similar group of industrialists. As his agenda noted, such a group would "advise of [the] likelihood of other nations repeating what our industry has done"—that is, whether other nations could build the vast, innovative industrial plant necessary to produce atomic bombs.

That May Monday morning the committee received copies of Bush's and Conant's September 30, 1944, memorandum to Stimson, the discussion framed on Bohr's ideas of the free exchange of scientific information and inspection not only of laboratories throughout the world but also of military installations. Bush promptly hedged his commitment to so open a world:

> I . . . said that while we made the memorandum very explicit, that it certainly did not indicate that we were irrevocably committed to any definite line of action but rather felt that we ought to express our ideas early in order that there might be discussion as [a] result of which we might indeed change our thoughts as we studied into the subject further, and I said also that we would undoubtedly write the memorandum a little differently today due to the lapse of time since last September.

At the end of the meeting Byrnes took his copy along and studied it with interest.

The Secretary of State-designate was learning fast. When the Interim Committee met again on Friday, May 18, with Groves sitting in, Byrnes brought up the Bush-Conant memorandum as soon as draft press releases announcing the dropping of the first atomic bomb on Japan had been reviewed. It was Bush's turn to be absent; Conant passed along the news:

> Mr. Byrnes spent considerable time discussing our memorandum of last fall, which he had read carefully and with which he was much impressed. It apparently stimulated his thinking (which was all that we had originally desired I imagine). He was particularly impressed with our statement that the Russians might catch up in three to four years. This premise was violently opposed by the General [i.e., Groves], who felt that twenty years was a much better figure. . . . The General is basing his long estimate on a very poor view of Russian ability, which I think is a highly unsafe assumption. . . .
>
> There was some discussion about the implications of a time interval as short as four years and various international problems were discussed, particularly the question of whether or not the President should tell the Russians of the existence of the weapon after the July test.

Bohr's proposal to enlist the Soviet Union in discussions before the atomic bomb became a reality here slips to the question of whether or not to tell the Soviets the bare facts after the first bomb had been tested but before the second was dropped on Japan. Byrnes thought the answer to that question might depend on how quickly the USSR could duplicate

the American accomplishment. The Interim Committee's recording secretary, 2nd Lieutenant R. Gordon Arneson, remembered after the war
of this confrontation that "Mr. Byrnes felt that this point was a very
important one." The veteran of House and Senate cloakrooms was at
least as concerned as Henry Stimson to extract a quid pro quo for any
exchange of information, as Conant's next comment to Bush demonstrates:

> This question [i.e., whether or not to tell the Russians about the atomic
> bomb before using it on Japan] led to the review of the Quebec Agreement
> which was shown once more to Mr. Byrnes. He asked the General what
> we had got in exchange, and the General replied only the arrangements
> controlling the Belgium-Congo [sic]. . . . Mr. Byrnes made short work of
> this line of argument.

The Quebec Agreement of 1943 renewed the partnership of the
United States and Great Britain in the nuclear enterprise; Groves was
justifying it as an exchange for British help in securing the Union Min-
ière's agreement to sell the two nations all its uranium ore. The British-
American relation was built on deeper foundations than that, and Conant moved quickly to limit the damage of Groves' blunder:

> Some of us then pointed out the historic background and [that] our con-
> nection with England flowed from the original agreement as to the complete
> exchange of scientific information. . . . I can foresee a great deal of trouble
> on this front. It was interesting that Mr. Byrnes felt that Congress would
> be most curious about this phase of the matter.

If Byrnes had begun his service on the Interim Committee respecting
the men who had carried the Manhattan Project forward, he must have
conceived less respect for them now. Both Stimson and Bush, Conant
told Byrnes, had talked to Churchill in Quebec. If, as it seemed, they
could be conned by the British into giving away the secrets of the bomb—
whatever Byrnes imagined those might be—for the price of a few tons
of uranium ore, how much was their judgment worth? Why give away
something so stupendous as the bomb unless you got something equally
stupendous in return? Byrnes believed international relations worked
like domestic politics. The bomb was power, newly minted, and power
was to politics as money was to banking, a medium of enriching exchange. Only naïfs and fools gave it away. . . .

CHAPTER 11

By the Bomb's Early Light: American Thought and Culture at the Dawn of the Atomic Age

Paul Boyer

The nuclear age began in 1945, with the detonation of the bombs over Hiroshima and Nagasaki, and no one has figured out a satisfactory means to bring us peacefully into a postnuclear age. Pandora's box is not easily closed. Even if the world's nuclear powers amicably agreed to total nuclear disarmament, nuclear knowledge could not be rendered unknown by simple treaty. We would still live under the potential nuclear threat of a renegade nation or a terrorist.

The nuclear threat has been a constant, and whether measured in terms of absolute numbers of warheads, or aggregate yield, or number of countries in possession of the necessary technology and fuel, the threat has varied only in the sense of growing distressingly larger over the intervening decades. Yet the psychological and cultural impact upon Americans of the knowledge that we live in a nuclear age has not grown correspondingly more visible over time. Concern about nuclear weapons has risen and ebbed in the United States in several cycles, obeying no neat pattern. In 1981, when Paul Boyer began to investigate the bomb's effects on American culture and consciousness, he was motivated partly by his desire to battle what seemed a profound public apathy toward the threat of nuclear war. But soon after he began, the American climate changed and public attention given to the nuclear threat increased noticeably. In his *By The Bomb's Early Light*, Boyer examines American society's earliest engagement with nuclear weapons, in the period 1945 to 1950.

Two selections from the book are presented here. The first is taken from a chapter entitled, "Dagwood to the Rescue: The Campaign to Promote the 'Peaceful Atom,'" which investigates the lack of sustained public interest in the atom bomb. In 1945 and 1946, the interest was strong, and public discussion of the bomb extensive. But by 1950, Boyer says, the bomb no longer preoccupied people; it had come to be passively accepted, and was infrequently mentioned if at all.

Boyer considers and rejects as insufficient one explanation of this decline in public interest: That people simply got tired of the subject after the endless discussions following August 1945. The subject got old, and was too morbid by its nature to sustain interest for more than a short while. People suffered from "sensory fatigue," and simply put the subject of atomic weapons out of mind. This, Boyer says, was only part of the explanation, however. The author presents an additional argument that public apathy toward nuclear danger was the product of propaganda campaigns launched by the federal government and the media.

The U.S. Atomic Energy Commission (AEC) was headed by a smooth, articulate chairman, David Lilienthal, who carefully kept the attention of reporters focused on the AEC's plans to develop nuclear-energy technology, which served to divert attention away from the AEC's nuclear weapons programs. Boyer suggests that this was an instance of conscious manipulation of the public because Lilienthal's own advisers were telling him privately at this time that atomic power was decades away from realization and was not given very high priority within the AEC.

The general public was given a positive view of atomic energy from the scientific popularizers, who explained scientific terms in reassuring everyday slang. "This new critter, the Atom" was introduced in relentlessly upbeat presentations, such as the AEC's "Atomic Energy Week," General Electric's comic book "Dagwood Splits the Atom," and CBS's hour-long 1947 documentary, "The Sunny Side of the Atom."

It would be unfair and anachronistic to criticize these materials for not anticipating events, such as the Three Mile Island and Chernobyl accidents, that would occur at nuclear power plants many years later. You will notice that Boyer does not use 1980s information to criticize the 1940s campaign to promote nuclear energy. Instead, he points to material in the 1940s that suggests that this campaign was used consciously to redirect public attention away from nuclear weapons and to secure public support for the AEC. Boyer maintains that David Lilienthal hid from public view the private concerns he had about nuclear weapons and the dangers they posed to world peace.

The second selection is taken from a chapter entitled "Social Science in the Breach," and shows us the early reactions to the atomic bomb of American professors—particularly in the fields of sociology, economics, psychology, and history. Boyer begins by looking at how the bomb caused discussion about how the curriculum would need to be changed in response. The nature of the bomb's fearsome power was seen to be so without historical precedent that some wondered whether the teaching of history could any longer be pertinent to questions of how best to respond to the bomb's new presence.

The author next shows that many social scientists saw a larger role for themselves outside of the classroom, in positions of political and social power. Only social scientists, they claimed, could help society live with the unbelievable power unleashed by the physical scientists. One management professor proposed that social scientists be given funding for a massive all-out research effort, on a scale equal to the Manhattan Project, which would enable social scientists to uncover the knowledge of human behavior necessary for peace in an atomic age. No such funding was forthcoming, however. The social scientists had to fight hard even to gain backdoor access to federal funds provided to a new National Science Foundation.

What would social scientists actually do if given real power to make decisions affecting society? Generalization is difficult, but a proposal offered by William F. Ogburn, a sociologist at the University of Chicago, certainly commands our interest. In 1946, Ogburn urged the complete and permanent relocation of all urban residents, moving everyone to hundreds of small towns dispersed throughout the countryside. Boyer interprets the proposal as an example of twin traditions in American social thought: hostility to the city and an impulse to have the state wield strong control over the urban masses.

We might read the Ogburn proposal differently, however, and instead emphasize how it is no more outlandish than other proposals for providing secure "civil defense" in the nuclear age. What would survive in American society after massive nuclear warfare? When the Pentagon or private think tanks have taken up this question more recently, and have tried to shape an answer to this question and to formulate detailed plans for postwar recovery, their efforts have been blasted by critics who charge that the military is "thinking the unthinkable" and deluding itself into believing that nuclear war is survivable in any meaningful sense. The same argument has been extended to our now-moribund civil defense system of designated shelters and emergency supplies. Before we debate how best to prepare for the possibility of nuclear holocaust, we must first settle the question of whether we should prepare at all.

DAGWOOD TO THE RESCUE: THE CAMPAIGN TO PROMOTE THE "PEACEFUL ATOM"

"The atom bomb, and all it means, does not appear to have sunk in at all. It has bounced off, as it were, or been mentally repelled as a tactless intruder." So wrote the British journalist Wyndham Lewis in 1949 after a visit to the United States. On the matter of the bomb, agreed Robert Payne in his 1949 book *Report on America*, people were behaving much as the medieval Christians who, though "conscious of an impending Day of Judgment, serenely abdicated their responsibilities to the church and to the state."

Whatever the bomb's deeper psychological effects, the surface evidence as the decade ended strongly supported such observations. In contrast to the bomb's massive initial impact, the climate had clearly altered dramatically. There were now fewer expressions of either cosmic despair or euphoric hope, fewer prescriptions for action, fewer pronouncements about the bomb's "larger meaning." As the open sense of urgency faded, the intense discourse of 1945 and 1946 diminished to scattered murmurs and faint echoes. The mood became one of dulled acquiescence. The bomb had come to stay. Represented visually, America's nuclear culture around 1950 would appear as a gray and largely deserted landscape.

To be sure, this cultural shift did not occur overnight. As early as February 1946, Raymond Gram Swing had sensed that the sharp edge of awareness was dulling. Already, he commented, the bomb "seems to have shrunk to something smaller . . . through the corrosion of familiarity." Bernard Baruch made the same point in a December 1946 United Nations speech: "Time is two-edged. It not only forces us nearer to our doom, if we do not save ourselves, but, even more horrendous, it habituates us to existing conditions which, by familiarity, seem less and less threatening. Once our minds have been conditioned to that sort of thinking, the keen edge of danger is blunted, and we are no longer able to see the dark chasm on the brink of which we stand."

By 1947, the shift was clearly evident. In June, summing up a series of recent polls, a public-opinion researcher observed: "On the whole . . . the threat of the bomb does not greatly preoccupy the people." At about the same time the number of entries on the bomb in the *Readers' Guide to Periodical Literature* began a sharp decline that continued with minor fluctuations until the mid-1950s. As J. Robert Oppenheimer remarked in a 1948 letter, there was a "surprising lack,

SOURCE: From *By the Bomb's Early Light* by Paul Boyer. Copyright © 1985 by Paul Boyer. Reprinted by permission of Pantheon Books, a Division of Random House, Inc.

both quantitatively and in discernment, in the public discussion of atomic energy."

Some saw this dramatic decline in overt concern about the bomb as a natural cultural reflex, a product of sensory fatigue. For a time after August 1945, the public's capacity to absorb books, articles, speeches, sermons, and discussions about the bomb had seemed boundless. But inevitably (or so this interpretation would suggest) other concerns and interests had reasserted themselves. Awesome in prospect, the atomic threat was simply less immediate than one's job, one's family, the cost of living, even the reviving rhythms of domestic politics and foreign affairs. As one public-opinion expert observed, there were distinct limits on people's capacity to sustain interest in any issue—even atomic war— "not intimately a part of their personal day-to-day preoccupations." Observed Norman Cousins following the Bikini tests: "After four bombs, the mystery dissolves into a pattern. . . . There is almost a standardization of catastrophe." Such explanations help us understand the lull in public attention and concern about the atomic bomb, but they hardly provide the full picture. The conscious manipulation of attitudes by government leaders and opinion-molders was involved as well.

The spontaneous initial surge of post-Hiroshima excitement over the atom's vast promise, quickly debunked by experts, had been short-lived. While it never disappeared entirely, particularly in reference to radioactive isotopes, a clear decline was evident by 1947. The end of the decade, however, brought a sharp quickening of attention to this theme in the media. Magazines once again blossomed with articles on the subject. *Newsweek* in 1949 enthusiastically reported "the new atomic story." The "frenzied wartime patchwork" that had produced the bomb had given way to "an orderly permanent enterprise" involving research and development for peacetime applications as well as weapons production. Under the enthusiastic headline "Plants, Payrolls, and Output Are Shooting Up," *U.S. News and World Report* reported that over sixty-five thousand workers were now employed in the atomic industry, and that the AEC was spending in excess of $600 million per year. While noting in passing that "improved atomic weapons" were being developed, this article left the distinct impression that these tremendous expenditures were primarily for peacetime research. Even weapons research, the magazine noted in 1950, was "certain to turn up scores of new civilian jobs for the atom."

Some expressed skepticism at the new media enthusiasm for "peacetime" atomic development. "Heartening as it is to know" of such programs, wrote Daniel Lang in the *New Yorker* in 1948, "the chilling fact remains that they are merely a by-product and that practically all the [AEC's] fissionable materials are allocated to nuclear weapons. To put it bluntly, more and better bombs are being made." Writing in *Christian Century* in 1949, two physicists noted that in the realm of the atom, "peaceful" research could readily have military applications. So long as the threat of war remained, they wrote, "atomic power will be of greatest benefit to man if it is confined to the sun and the stars."

Significantly, this new media blitz promoting peacetime uses of the atom came just as atomic-energy policymakers realized that such applications were increasingly remote. In a July 1947 report to the AEC, J. Robert Oppenheimer not only (in David Lilienthal's words) "discouraged hope of atomic power in any substantial way for decades, but . . . question[ed] whether it would ever be of consequence." ("Had quite a blow today," commented Lilienthal in his diary.) The AEC's general manager in the late 1940s, Carroll Wilson, later summed up the prevailing mood: "The power thing was pie in the sky, really. . . . All of the other priorities were higher than nuclear power."

Despite such behind-the-scenes gloom, media and public attention to the peacetime promise of atomic energy seemed only to grow as the 1940s ended. Far from being spontaneous, however, this revival was being consciously induced by government, corporate, and media manipulation. The aim was not primarily to publicize peacetime applications per se, but rather to create a more positive— or at least more acquiescent—overall public attitude toward atomic energy. From the beginning, of course, scientific popularizers had tried to convey basic information about nuclear physics in terms not only comprehensible but even appealing to the lay public, by introducing, for example, everyday allusions and reassuring analogies. One 1946 popularization said that the oscillating electrical fields used in isotope separation "make the ion beams dance in the very same way as the electrons dance in the amplifier tubes when some famous crooner sings on the radio." Another called electrons "electrical jitterbugs." *The Atomic Story*, by John W. Campbell, Jr., begins with a Cast of Characters bearing such cute names as "Proton: A plump, positive fellow." Still another early popularization featured illustrations by the young Maurice Sendak, including one picturing atoms as dancing men and women pairing off to form molecules.

A certain innocent exuberance—plus the desire of specialists to demystify their subject—pervades these early popularizations. In the later 1940s, such benign imagery was manipulated much more consciously to allay the public's atomic fears. Central to this effort was the Atomic Energy Commission and its articulate chairman. Often featured in the press as "Mr. Atom," David Lilienthal strove mightily to give a benevolent human face to a reality whose terrifying aspects never lay far beneath the surface. As we have seen, Lilienthal in the privacy of his journals, and occasionally more openly, betrayed deep anxiety over the darkening nuclear weapons situation. In public, at who knows what psychic cost, he usually maintained a determinedly upbeat tone in discussing what he sometimes called "this new critter, the Atom." In articles in the popular press and speeches he wrote himself and delivered in a

colloquial, extemporaneous style, Lilienthal (despite occasional caution-
ary notes) generally extolled the atom's peacetime promise and encouraged positive thinking. In a 1948 high-school graduation address of
almost strident optimism delivered at Gettysburg and broadcast nation-
ally, Lilienthal acknowledged that many opinion-makers were in "deepest despair" about the atomic bomb. He urged Americans, however, to
ignore their "predictions of dire and utter calamity." Every generation,
he said, had its pessimists convinced "the world was going to the dogs."
With "knowledge, love [and] faith," he concluded, the atomic age could
be "one of the blessed periods of all human history." He sought to ease
fears over the bomb itself in insisting—a theme picked up by *U.S. News*
and other publications—that research on atomic weapons and development of the atom's "beneficent and creative" potential were "virtually
an identical process: two sides of the same coin." The former would
inevitably promote the latter.

    Another favorite Lilienthal theme was that atomic energy was sim-
ply a form of solar energy—and who was afraid of the sun? As he put
it in another high-school commencement speech, this one in Crawfords-
ville, Indiana: "I suppose there is nothing of a physical nature that is
more friendly to man, or more necessary to his well-being than the sun.
From the sun you and I get . . . the energy that gives life and sustains
life, the energy that builds skyscrapers and churches, that writes poems
and symphonies." Yet this benevolent, life-giving sun, Lilienthal concluded triumphantly, was nothing but "a huge atomic-energy factory."

    Reflecting Lilienthal's upbeat approach, the AEC directed its atten-
tion in the late 1940s—when bomb-making was by far its highest
priority—to projecting what its official historian called "a peaceful, civilian image." A Division of Public and Technical Information was set
up in October 1947 to build favorable public relations through work with
"the press, the radio, schools, organized groups and others." In a pilot
project that autumn, the AEC organized an Atomic Energy Week in
Hyattsville, Maryland, just outside Washington. Sidewalk informational
tables were set up, eye-catching exhibits displayed, and ministers and
newspaper editors urged to discuss atomic energy in their sermons and
editorials. High-school students put on a play featuring such characters
as Miss Molecule and Mr. Atomic Energy. Grade-school children were
given simplified talks on the atom's peacetime potential and the Baruch
plan. The county's seven hundred teachers were lectured by atomic
scientists. The whole endeavor was praised by the *New York Times* as "a
broad-scale public seminar in just what the atom is all about."

    Newspaper accounts of Hyattsville's Atomic Energy Week pre-
sented it as a grass-roots undertaking, but the AEC's role became evident
in 1948 when communities across the country began organizing similar
events. As in Hyattsville, sponsorship by local civic groups was arranged, exhibits were set up, and lectures were given for different age

groups. An exhibit was supplied to the Iowa State Fair; Lilienthal himself inaugurated a show in Cincinnati. The overall aim, said one AEC official, was to demonstrate that "atomic energy is already at work for *good*" and to prepare Americans to "assume their atomic-age responsibilities." In 1947 and 1948, the AEC estimated, four million visitors attended exhibits sponsored by the AEC or its corporate contractors.

Prominent in this effort was the Brookhaven National Laboratory, a Long Island facility jointly funded and administered by the AEC and nine large eastern universities. Much of the AEC's limited nonmilitary research was centered at Brookhaven, and its staff frequently spoke to public gatherings "trying to prove, against heavy odds, that atomic energy has its attractive side" (as the skeptical Daniel Lang put it). Brookhaven's public-relations office assembled two traveling exhibits featuring movies, audiovisual displays, and live demonstrations. Exhibitions were held in a number of cities and at the American Museum of Natural History in New York. Reporting a Brookhaven road-show presentation in Stamford, Connecticut, under the title "Main Street Meets the Atom," *Science Illustrated* described the "sober fascination" of adult visitors and the "delighted" reaction of children to such exhibits as a Van De Graaff generator and an "atomic pinball machine"—an atomic-pile simulation complete with gong and blinking lights.

The high point of this exercise in government-inspired positive thinking came in the summer of 1948 with Man and the Atom, a month-long exhibit in New York's Central Park. The show was sponsored by the AEC; its major corporate contractors for nuclear power development, General Electric and Westinghouse; and the New York Committee on Atomic Information—an umbrella organization of local backers. The exhibit's centerpiece was a Theater of Atoms, sponsored by the Westinghouse Corporation. The theater featured such eye-catching exhibits as a "real radiation detector," a "chain reaction" (of sixty mousetraps), and a model of an atomic nucleus which, according to a Westinghouse public-relations spokesman, resembled "a futuristic Christmas tree" and exploded harmlessly with "a blinding flash! and ear-splitting crash!"

Visitors to the General Electric exhibit received free copies of *Dagwood Splits the Atom*, a colorful comic book produced by King Features Syndicate in consultation with the AEC, in which Mandrake the Magician reduces Dagwood and Blondie to molecule size to unfold the wonders of the atom to them while an audience including Popeye and Maggie and Jiggs looks on. (General Groves himself, in yet another of the high-level policy decisions of his career, chose Dagwood as the central character.) Over 250,000 copies were distributed, leading GE to order a further printing of several million.

As the AEC's "public information" campaign gained momentum, it shifted subtly in emphasis. The Hyattsville Atomic Energy Week of 1947, though generally upbeat, did include information on the bomb's destructive power and the need for international control. In the exhibits of 1948, this emphasis largely disappeared, and the focus was almost entirely on soothing public fears. At the Man and the Atom show in Central Park, apart from a single exhibit showing the effects of radiation on blood cells and urging people to give blood in preparation for "any atomic-bomb emergencies," the emphasis was exclusively on stimulating positive attitudes toward atomic energy. In *Dagwood Splits the Atom,* when Mandrake sets off a chain reaction (as silly Dagwood rushes off in a panic shouting "Blondie!"), the "BANG!" in the center of the picture is balanced by drawings of a power plant, a factory, a grain field, a medical lab, and an atomic ship—the many uses of the peaceful atom. Even if most of the peaceful uses featured in the exhibit either lay far in the future or had already been discredited, Westinghouse's public-relations spokesman disingenuously insisted, surely this was "a better note to strike . . . than emphasizing the destruction of bombs."

That the goal of all these Atomic Energy Weeks, Man and the Atom shows, and related activities was less educational than therapeutic and propagandistic emerged in the January 1949 *Journal of Educational Sociology,* a special issue devoted entirely to the previous year's blitz of semiofficial atomic-energy exhibits. The editor emphasized the importance of publicizing the "magnificent story" of the atom. General Groves urged pride in the Manhattan Project and then moved briskly onward:

> Enough of the past—let's talk about the present and future. . . . Much that has been written about atomic energy has inspired fear and confusion. . . . This is not a healthy state of affairs. Atomic energy must be explained. The average American likes new scientific devices. He must learn that nuclear energy, like fire and electricity, can be a good and useful servant.

A Pentagon spokesman complained that too many people regarded atomic energy as "black magic" and that most writers on the subject— "as ill at ease . . . as a Victorian concert audience suddenly subjected to modern jazz"—were merely projecting their own worries. We must, he insisted, "absorb elementary nuclear science into our folklore as soon as possible."

The editor of *Popular Science,* in a similar vein, noted that in 1945–1947 many erroneous and ill-informed notions had been implanted in the public mind, but added that "those days, fortunately, are over now." Scientists, he said, were now inviting journalists into the laboratory to explain "their creed and their work"; Nobel laureates were making themselves "nearly as accessible as politicians"; and the AEC was cooperating "with responsible writers and artists as much as conditions and legal restrictions permit."

Though spearheaded by the AEC, this governmental effort to "domesticate" the atom and play up its peacetime uses took many forms. A 1949 State Department film carried the message to foreign lands. "We have found an increasing need for counteracting some of the effects abroad of the well-known destructive and wartime potentialities of atomic energy," wrote a State Department public-relations staff member. "This short film would be designed to open up some of the peacetime vistas."

Educators played a central role as well. The AEC's public-information office worked closely with the U.S. Office of Education and the American Textbook Publishers Institute, for whom it prepared a *Sourcebook on Atomic Energy*. It also cooperated with the National Education Association in the writing of *Operation Atomic Vision* (1948), a ninety-five page high-school study unit. This handbook urged students to promote interest in atomic energy through discussion groups, lecture series, school projects, letters to the editor, and Community Atomic Energy Councils. The danger of atomic war was discussed, but the clear intent was to supplant this negative image:

> Atomic Energy! What do you think of when you hear these words? The chances are that these words call up in your mind thoughts of war, destruction, the atomic bomb. This is not strange for, after all, the press and radio and prominent people have emphasized the great hazard of war and the A-bomb. . . . You may even wish to bury your head in the sand and resign yourself to fate. But there is a much brighter, a much more constructive, and a much more thrilling side of the atomic energy picture. If we look long enough and hard enough at this side of the picture, we might be able to see a world free from war, strife, poverty, and sickness; a world of hope and of great possibilities for human welfare. . . . Why not keep the bright side of the atomic energy picture in the center of our attention?

Operation Atomic Vision proceeds to picture not only the wonders of isotopes, but an atomic future reminiscent of the most euphoric effusions of the immediate post-Hiroshima period:

> You may live to drive a plastic car powered by an atomic engine and reside in a completely air-conditioned plastic house. Food will be cheap and abundant everywhere in the world. . . . No one will need to work long hours. There will be much leisure and a network of large recreational areas will cover the country, if not the world.

In *The Challenge of Atomic Energy*, a curricular guide published by Teachers College of Columbia University as a supplement to *Operation Atomic Vision*, one finds (despite the apprehensions noted in the last chapter) equally glowing descriptions of an atomic future of boundless food resources, "unlimited, almost automatic, effortless production," and "a fantastic standard of living . . . that might very well abolish what we now call the 'deprived' class in our society." The aim of this curricular project, it was said, was to replace "irrational fears" with "improved attitude and action patterns in this area."

Jumping aboard the bandwagon, the nation's educational journals increasingly touted the atom's peacetime uses of 1948–1949. "Simply to go on repeating that the bomb is awful, that there is no defense against it, and that we must have world control, can only lead to panic and fatalism," said an Iowa teachers' journal in September 1948. "The American people need to be given a new approach to the subject . . . based on calm acceptance of atomic energy as part of our life and one of enormous significance other than for war." "The job of the teacher is not to scare the daylights out of people by regaling them with the horrors of atomic destruction," agreed a University of Illinois education professor. Their emphasis, he said, should be "on the positive aspects of atomic energy control in the service of mankind."

The scientific popularizers joined in as well. In *Energy Unlimited* (1948), Harry N. Davis wrote: "The outlook of this presentation is optimistic. Too many people are counting in advance the possible casualties of future war with the new weapons of atomic power . . . and falling into hopeless despair." The best way to assure peace, he said, was to "keep our eyes on the wonderful goals that are almost within our grasp through the positive use of the forces science has unlocked." While offering a cautious assessment of the possibilities for atomic power, Davis wrote enthusiastically of the medical potential of radioactive isotopes.

The radio networks, too, cooperated with the governmental effort to mute "excessive" fear of the bomb. In 1946, a radio had offered such somber fare as Hersey's *Hiroshima* and realistic dramatizations of an atomic attack. In June 1947, by contrast, CBS broadcast an hour-long documentary called "The Sunny Side of the Atom" designed, according to a spokesman, to counteract "the 'scare' approach to atomic education." With Agnes Moorhead in the role of a peripatetic narrator, this pseudo-documentary begins with a "visit" to Oak Ridge, where radioactive isotopes were being produced. The propaganda was not subtle. The looming secret facility at first seems sinister, but "above the eerie whistling of the smokeless stack, I heard the birds, singing." The scene next shifts to a doctor's office, where, thanks to isotopes, a golf pro learns that his foot won't have to be amputated. "You may be out there shooting in the seventies again before too much longer," the kindly doctor tells him. (Only the most attentive listener would have realized that the isotopes did not actually cure the patient, but only facilitated the diagnosis.) On to a great medical research facility, where researchers seek "a cure for cancer with the aid of isotopes."

Shifting to the Southwest, "The Sunny Side of the Atom" next dramatizes the role of isotopes in seeking out residual oil in abandoned wells. Right on cue, a prospector exclaims, "Unless I miss my guess, I think we've struck oil." Ruminates a local: "Guess folks around here won't feel so bad about that atom bomb when they hear about this." (Music "up humorously," directs the script.) After a final visit to an "atom farm" where vast isotope research projects are underway, Agnes Moorhead soliloquizes:

> But that wasn't all, nearly all. The rest of the story was not for our eyes— it was for our hearts and our hopes. It was the personal visions of the hundreds of men and women—scientists, engineers, technicians—who spoke to me frankly at all of our great atomic research centers. They are the custodians of the infant science. And they had bold ideas—about how it can grow and what it can mean tomorrow—for peace and plenty, for health and better living everywhere. If the world will only give them the chance—and the time. (*Music stabs climactically, but quickly broadens and segues into the intimations of a rosy dream.*)

Overwhelmed by all she has seen and heard, the narrator lapses into a trancelike reverie:

> I saw a great light, more luminous than the sun, flooding out over the darkness of the earth. Its rays were lighting up the innermost recesses of life, searching out secrets never revealed by the light of the sun. . . . I saw all men standing straight and tall and confident, facing, without fear, their future, urged forward by a new hope, by the infinite wonder and possibility of a new life.

After a final fleeting reference to the horror of atomic war, optimism triumphs:

> "Everything I have seen, everything I have heard, everything I have felt has given me this faith: We are bigger than the atom, and if we face the future boldly, we will enter a world made bright by the sunny side of the atom."

The psychological intent of this multifaceted campaign of "atomic education" in the late 1940s is quite clear: to implant in the public mind an image of atomic energy associated with health, happiness, and prosperity rather than destruction. Eager to document this shift, CBS's research department assembled a group of test subjects to listen to "The Sunny Side of the Atom." A polygraph hookup measured their moment-by-moment emotional responses, red and green buttons allowed them to indicate approval or disapproval of various segments, and a follow-up questionnaire was administered. "The over-all emotional effect," the researchers reported, "was to lessen fear of atomic energy, with 46 percent of the group less fearful than before the sample broadcast, 3 percent more fearful, and 51 percent with fears unchanged."

A similar improvement in attitude was found among visitors to New

York City's Man and the Atom exhibit of 1948. Under the direction of Lillian Wald Kay, a psychologist at New York University, entrance and exit polls were administered to two thousand visitors. When asked to list some uses of atomic energy, only 33 percent of those leaving the exhibit mentioned war, in contrast to 42 percent of those in the entrance poll; while the number mentioning medicine, power, and industrial and agricultural applications increased markedly. Asked to choose from a list of words the one best describing their feelings about atomic energy, 34 percent of the exit-poll subjects selected "Hope," compared to the 27 percent in the entrance poll.

This systematic effort by opinion-molders in government, education, and the media to reshape public attitudes toward atomic energy involved a conscious repudiation of the "fear" strategy of the scientists' movement. This fact was heavily underscored by Lillian Wald Kay in two articles published in professional journals in 1949. The early post-Hiroshima media preoccupation with "the problem of the bomb and its control," she began, had encouraged Americans to think of atomic energy solely as a frightening, destructive power, not as "a possible force for good." As a result, "atomic energy" had become firmly anchored in people's minds with thoughts of "war and the fear of war." This, in turn, had led to "a gradual decrease in enthusiasm" for it and a lack of "clarity" in public perceptions of its potential benefits. Illustrating this progression, Kay cited an interview subject who reported that his initial, spontaneous reaction to the news of Hiroshima had been positive, as he had thought of "the great change in human living atomic energy would bring." But then, fed by a wave of "propaganda" about atomic war, fear had pushed aside hope. Citing public-opinion data, Kay noted that between 1945 and 1947, the percentage of Americans who considered the development of the atomic bomb a "bad thing" had more than doubled. These postwar cultural and attitudinal trends, Kay concluded, posed a clear challenge to "psychologists and educators" who wished to help a worried public "prepare for [the atom's] promise as well as to face its problems." The ambivalence revealed in the polls, she said, made clear that Americans were "ready for well-directed attempts to change their opinions."

Concluding one article on a practical note, Kay reported the discussions of a small "brainstorming" group she had assembled to consider ways of publicizing the positive side of the atomic story. The suggestions included publicizing the pro-atomic-energy views of celebrities and dramatizing the atom's benefits. "Show them a train run by atomic power and they won't worry about the atomic bomb," said one participant; emphasize the medical benefits, suggested another, since "so many people worry about cancer." Whatever the specific strategies, Kay's underlying message was clear: experts must devote their energies to "influencing perception away from the bombs and back to energy." The challenge was no mere "interesting theoretical problem" but "the most vital of our times."

The motives behind this systematic effort to create a more positive public perception of atomic energy were certainly mixed. Many of the educators, psychologists, editors, and radio executives who implemented it doubtless believed they were performing a patriotic public service. For David Lilienthal and others of the AEC, the rhetorical insistence on the atom's exciting peacetime potential helped mask the lack of actual progress in this area. Lilienthal's constant round of speech-making served to build a public constituency for the AEC and also protect him from congressional critics, some of whose personal hostility went back to his days as head of the "socialistic" Tennessee Valley Authority. Theoretically committed to the principle of civilian control of atomic energy, Lilienthal needed to hold out at least the prospect of "peaceful uses" to sustain the rationale for that principle, even as the AEC in practice increasingly became a creature of the military.

Above all, this hoopla for "the sunny side of the atom" must be seen in the context of a deepening Cold War, in which America's military planners were placing increasing strategic reliance on the nation's nuclear weapons program—a rapidly expanding program shrouded in secrecy. With the nuclear arms race entering a dangerous new phase, the public was urged to contemplate the vast and friendly benefits of the peacetime atom. In February 1950, following a meeting with President Truman two weeks after Truman's go-ahead on building the hydrogen bomb, Lilienthal recorded in his diary the president's full agreement "that my theme of Atoms for Peace is just what the country needs."

SOCIAL SCIENCE INTO THE BREACH

Lewis Paul Todd was a young instructor at a Connecticut teachers' college when he heard the news of Hiroshima. Within hours, he began an impassioned essay published a few weeks later in a journal for high-school social-studies teachers. The news of August 6, he wrote, had made vividly apparent the chasm "between man's ability to solve the problems of the physical universe and his utter inability to solve social problems." The imperative response, Todd continued, must be to match the unleashing of atomic energy with "a revolution of equal force in the world of human relations." In an age of atomic bombs every other activity must be subordinated to "the job of *social engineering*."

Todd was far from alone in this conclusion. Indeed, it crops up with almost hypnotic frequency in post-Hiroshima social commentary. Hiroshima, commented Lyman Bryson in a CBS radio broadcast soon after the event, underscored the urgency of applying "the same kind of thinking in managing men as we can show in managing atoms." It was high time, said the *New Republic* on August 27, for "the science of human personality and . . . society" to receive the same urgent attention "hitherto devoted to discovering the secrets of matter." The atomic scientists had done their job, added the *Antioch Review* in December; society now awaited "the guidance of the social scientists."

The philosopher John Dewey, who for decades had urged the application of scientific method to social problems, renewed his plea in the aftermath of Hiroshima. Writing early in 1946, Dewey lamented the "tragically one-sided development of knowledge" that had left man's social knowledge "in an infantile state" relative to his "discoveries, inventions, and technologies." "Natural science has far outstripped social science," agreed Harvard historian Sidney B. Fay in his 1946 presidential address before the American Historical Association. "We have discovered how to split the atom, but not how to make sure it will be used for the improvement and not the destruction of mankind."

What was to be done? Many commentators insisted that the teaching of the social sciences, and especially history, must be radically revised to reflect atomic-age realities. The ferocity of World War II, culminating at Hiroshima and Nagasaki, wrote E. B. White in the *New Yorker* a few weeks after the war's end, underscored the irrelevance, if not the actual meretriciousness, of the traditional patriotic classroom approach to history, with its focus on nation-states, and their wars and conflicts. "We take pains to educate our children at an early age in the rituals and mysteries of the nation," White observed, " . . . but lately the most conspicuous activity of nations has been the blowing of each other up, and an observant child might reasonably ask whether he is pledging allegiance to a flag or to a shroud." In the same vein, Princeton's Christian Gauss paused in his *American Scholar* attack on the parochialism of literature teachers to take on the historical profession as well. In a misguided attempt to be "scientific," he charged, historians had "refrain[ed] from passing moral or 'value' judgments" and emphasized disembodied "historic forces" over individual choice and responsibility. Yet for all the profession's pretensions to objectivity, he went on, echoing E. B. White, textbook history remained narrowly nationalistic at a time when global thinking had become essential. Gauss even implied that all history prior to August 6, 1945, had been rendered irrelevant by the atomic bomb. Would "even the most accurate and dispassionate account of every stage that led to the invention and dropping of the bomb," he asked, "help us very much in reaching a wise solution of the problem with which science has confronted us?" If historians were to play any meaningful role in the atomic era, he concluded, they would have to rethink radically the nature of their enterprise.

Some social-science teachers were themselves reaching the same conclusion. These subjects would have to undergo a thorough "reconstructing and revitalizing," warned Alonzo B. May of the University of Denver, to be of much value in a society transformed by atomic energy. If a comprehensive revision of "the social studies curriculum . . . was necessary a year ago," agreed the headmaster of a New Hampshire high school early in 1946, "it is even more so today." Though the bomb seemed overwhelming, he insisted, educators in the nuclear era must "strive to break down old cultural fears and shibboleths and encourage a more fearless questioning of the social order . . . in the hope that a new generation may be adaptable to social change." In an age of atomic bombs, declared Lewis Paul Todd, "the social sciences must become the keystone . . . of public education. And . . . the increased time we devote to the study of human relations must be taken from the physical sciences." Todd insisted that if the social-studies curriculum were to prove equal to the crisis created by the bomb, it must be organized around a "directing moral principle" to provide the rising generation the ethical orientation it would urgently need in the atomic era.

While social-studies teachers pondered the pedagogical implications of the atomic bomb, others spoke expansively of a vastly enlarged public role for social science. In his 1939 book *Knowledge for What?* sociologist Robert S. Lynd had called for an activist, engaged social science, and in 1945 Lynd's summons seemed more germane than ever. Within forty-eight hours of the Hiroshima news, a group of five prominent social scientists including sociologist Talcott Parsons and political economist Lincoln Gordon submitted a lengthy letter to the *Washington Post* asserting that world peace had become society's most urgent challenge in light of "the startling news of the atomic bomb" and that in this task the social sciences had a vital role to play. Cooperative human intelligence could solve "human problems as well as . . . those of atomic physics," they insisted, provided it proceeded according to a method, and "this method must be social science." They urged a high-level study to "explore the needs which social sciences must fill in a world equipped for suicide, and the means to put social science brains in harness." The planning advocate Stuart Chase quickly caught the mood. "As the first item on the educational agenda," Chase wrote late in 1945, "I respectfully suggest that another two billion dollars be allocated, this time to the *social scientists.* An equally urgent directive should go along with it: [to] show us how to live with the unbelievable power the physical scientists have loosed upon us."

One of the strongest expressions of this exalted view came from John S. Perkins, professor of management at Boston University. Sitting at his typewriter on the evening of V-J Day, Perkins wrote an article that soon appeared in *School and Society,* "Where Is the Social Sciences' Atomic

Bomb?" "Man's behavior and that of his institutions must now be harnessed just as the behavior of atoms has been harnessed. This is the result the social scientists must produce. This is their atomic bomb. Can they come through?" The social scientists would succeed at this task, Perkins went on, only if they emulated the organizational genius of the atomic scientists in mobilizing "the ideas of men all over the world" in a coordinated attack on an urgent problem. "Only by the same kind of . . . concerted, all-out effort where past procedures are thrown by the board can the social scientists produce their atomic bomb. "Swept up in the excitement of his vision, Perkins summoned social scientists to something akin to a religious crusade. "The top minds of the world must be commandeered," he said, to "harness the complex and powerful aspirations, actions, and attitudes of men and channel them into the ways of peace. This is their Manhattan District Project. Upon their success now rests the fate of all mankind." Such a crusade, he went on, would surely win for social scientists "respect in the halls of government and in the minds of the people—respect which has been missing to a notable extent in the past. . . . The physical scientists . . . have deservedly won the praise which a still incredulous world is pouring out to them. . . . Now it is the social scientists' turn."

A similarly expansive vision of the social scientists' role in the atomic era came in a 1946 *New York Times Magazine* article, "We Need a New Kind of Leader," by psychologist A. M. Meerloo. Man's only hope in the atomic era, he said, was "scientific long-range social planning," and this would require experts: "We are desperately in need of social scientists today—instead of the atomic physicists who try to give social advice." Only "a united social science—embracing economics, sociology, and psychology," and led by "free intellectuals with original thoughts, who will not bend to authority" would be adequate to the atomic crisis. The United Nations should at once set up "a scientific social board," Meerloo concluded, "to lead the world into sane and workable channels of peace."

Though rarely with such soaring rhetoric, many social scientists in this post-Hiroshima period embraced the view that they possessed knowledge and expertise essential to mankind's survival. With humanity "haunted by the uneasy fear that an unconquerable monster has been released," a sociologist stated in *Science* early in 1946, social science must devise "an economic, social, and political organization" that would promote "human values, health, love, emotional adjustment, and security." At a moment when the bomb had made the threat of extinction real and immediate, social psychologist Kenneth Clark told the National Committee for Mental Hygiene in 1947, "the retardation in the development of the social sciences" must be urgently addressed.

Questions of status, prestige, and public acceptance were never far from the minds of sociologists as they speculated on the role of their discipline in the age of the atom, but something more concrete was also involved. The issue of federal support for social-science research was of immediate concern to sociologists in the 1945–1950 period. During the war, scientific research on military projects had been funded through the Office of Scientific Research and Development, headed by Vannevar Bush of MIT. (The atomic bomb project itself was under OSRD until 1943, when it was transferred to the army.) As the war drew to a close, Bush began to campaign for a continued flow of federal dollars to scientific research in the postwar period. In *Science: The Endless Frontier* (1945), an OSRD report to President Roosevelt, Bush called for the creation of a new federal agency, the National Science Foundation, as a conduit for such support. In the early postwar months—and intermittently for the next five years—this proposal generated heated debate in Washington and in the press.

Would the proposed foundation support the social as well as the natural sciences? Psychologists, sociologists, economists, and other social scientists had engaged in federally funded research through President Herbert Hoover's Research Committee on Social Trends (1929) as well as various new Deal and World War II agencies, and many now insisted that social-science research be included in any postwar program of federal support of science. Vannevar Bush disagreed, however, and the original NSF bill (introduced in July 1945 by Senator Warren Magnuson) reflected his views. Influential social scientists at once raised objections. The Social Science Research Council—established in 1923 by the professional societies of the various social sciences to promote interdisciplinary cooperation, more rigorous research standards, and a larger public role for these disciplines—took the lead in the campaign to incorporate the social sciences in the proposed legislation. In October, Senator Harley Kilgore of West Virginia introduced an alternative NSF bill that did, indeed, include the social sciences.

In the early post-Hiroshima years, then, discussion of social science's role in the atomic age frequently became enmeshed in the debate over the scope of the National Science Foundation. Not only leading social scientists, but also influential natural scientists, vocally supported the Kilgore bill in the fall of 1945. Forming the Committee for a National Science Foundation, astronomer Harlow Shapley and chemist Harold Urey secured the signatures of hundreds of scientists, including Albert Einstein, Enrico Fermi, and other notables, to a petition calling for the inclusion of "all fields of fundamental scientific inquiry relevant to national interest without arbitrary exclusion of any area." A poll of some six hundred leading members of the American Association for the Advancement of Science revealed that nearly 70 percent favored federal funding of social-science research. President Truman let it be known that he supported the Kilgore bill.

The issue came to a head in October 1945, when a subcommittee of the Senate Committee on Military Affairs conducted hearings on the various NSF bills then before the Congress. While a few witnesses, including the spokesman for the American Chemical Society, explicitly opposed inclusion of the social sciences, most of those who addressed the matter, including J. Robert Oppenheimer, President Karl Compton of MIT, President James Conant of Harvard, and, of course, the big guns of the Social Science Research Council, favored such inclusion.

Many of these witnesses couched their support for federal funding of social-science research explicitly in terms of the social crisis the atomic bomb had created. "Ignorance of the science of humanity will lead us inevitably to destruction," declared Commerce Secretary Henry A. Wallace. "Our great problems . . . are not the problems of the natural sciences," added the Princeton physicist Henry D. Smyth. "They are the problems of the social sciences, and of politics and of ethics, if you like." Federally funded research that simply made possible "bigger and better atomic bombs" without helping society learn "how to live without using these new weapons," said historian John Milton Potter, president of Hobart and William Smith Colleges, would betray "the aspirations of most Americans."

This theme was stressed, too, in the extended plea for federal funding of social-science research made in 1946–1947 by Talcott Parsons. The atomic bomb, wrote Parsons in the December 1946 *American Sociological Review,* had dramatically underscored "the potentialities of modern scientific technology for destruction and disruption of social life," and a social-science response was imperative. While the physical scientists' "enormous popular prestige" gave an "oracular" quality to "their pronouncements on almost any subject, whether or not it falls within their field of special competence," he said, social scientists were in fact the best qualified to confront the social problems science had created. Acknowledging social science's "distressing history of ineffectuality" in contrast to the physical sciences, he insisted that in several areas—organization theory, public opinion, demography, the business cycle—it could already "deliver results of first-rate practical importance" and that on all fronts it was rapidly progressing toward full scientific legitimacy.

Pursuing the theme in the January 1947 *Bulletin of the Atomic Scientists,* Parsons developed his argument by stressing the unity of science and rejecting any definition that would exclude the study of society. "It simply is not possible to draw sharp, clear-cut lines between the natural and the social sciences," he insisted, citing medicine as an example. "Each side needs the contributions of the others, and many of the most important problems fall across the line." Public funding of social-science research, he concluded, would be a symbolically important acknowledgment that "science" was "not a mere isolated technical tool in modern western society" but a mode of thinking whose "roots penetrate to the deepest levels of our cultural and moral motivations."

A few critics, however, challenged the assertion that social-science research represented America's greatest hope in the nuclear age. As early as November 1945, the director for social sciences of the Rockefeller Foundation deplored the fact that "many intelligent people" seemed convinced that the problems of the atomic age could be solved by some "mechanistic solution" offered by social scientists. No single, comprehensive "answer" from social science, he said, would eliminate the continuing threat posed by the bomb. Another skeptic, at least in the privacy of his journal, was David E. Lilienthal. In a series of annoyed entries from 1946 to 1948, Lilienthal fulminated against the widespread view that all would be well as soon as social science could "catch up" with the natural sciences. Social and political change did not come about through any "process of pure reasoning" by social scientists, he wrote. Indeed, it was "not affected by social scientists at all" but by technological and scientific developments. The social scientists "merely record changes that have occurred, or predict ones to come, or analyze the facts, etc." It was therefore vain, Lilienthal insisted, to look for an answer to the atomic dilemma through the accumulation of social-science data.

Religious writers were also particularly acute in criticizing the hopes some were investing in the social sciences. Writing in 1948, L. Harold DeWolf of the Boston University School of Theology rejected the widespread assumption that mankind's plight was "due to a cultural lag in which the physical sciences have too far outstripped the social sciences." He also challenged the belief (implicit in this assumption) that while technological and scientific advances had both positive and negative potential, social-science research would be used only in beneficent ways. Pointing out that Hitler and Stalin had perfected sophisticated techniques of social control, DeWolf insisted that "the power given by the study of man," no less than the power given by nuclear physics, "can be used for either good or ill." Volumes of social research might accumulate on the shelves, but the question of the ends and purposes of that knowledge would remain, and social science could never in itself provide the answer.

Meanwhile, as the debate continued, the proposal for a National Science Foundation had become involved in numerous other controversies and political crosscurrents, and not until the spring of 1950 was the enabling legislation finally passed. As ultimately set up, the NSF was mandated to support basic research in the "mathematical, physical, medical, biological, [and] engineering" sciences, though a loophole reference to "other sciences" opened the door to the social sciences at least a crack.

Despite the scattered notes of caution, the conviction that social science could resolve the atomic dilemma found many adherents in these years. Indeed, a few social scientists went beyond broad generalizations and actually tried to apply their expertise to specific atomic-bomb-related issues. The results were not encouraging. In his 1946 essay on "Sociology and the Atom," for example, William F. Ogburn devoted considerable attention to a single issue: the relocation of city dwellers as a response to the danger of atomic attack. Ogburn simply began with the premise that such an urban dispersal project was necessary and desirable. Downplaying the daunting practical difficulties of such an undertaking, as well as the warnings of some that such an exercise in forced relocation would turn America into a police state, Ogburn discussed this bizarre idea as though it were, in fact, a serious policy option. Tapping into the nightmarish visions of urban destruction that had swept over America, Ogburn built a case for breaking up the nation's cities into hundreds of small towns scattered across the countryside. Those who found such ideas "utterly impracticable," said Ogburn, should consider that an atomic war could wipe out "fifty million city dwellers in a few minutes." Citing his own institution as an example, he acknowledged that "to move the University of Chicago a hundred miles away" might seem "a very difficult task," but insisted that it must be viewed in the context of the fact that

> within twenty-five years an atomic bomb may melt down all the buildings now on campus and all the equipment, books and laboratories. . . . A crisis great enough to spur us to dismember our cities is almost sure to come sometime. But then it will be too late. Are we farsighted enough to act in advance?"

In an imposing display of analytical precision, Ogburn calculated that the cost of breaking up America's cities into a thousand smaller ones would be $250 billion, "less than the cost of the second World War to the United States and perhaps less than the cost of a third world war." He also introduced historical precedent to buttress the feasibility of his project:

> The Pueblo Indians once moved their cities from the plains, where they were the prey of their warlike enemies, and set them in caves scooped out of canyon walls high above the river. This was a task as difficult for them, perhaps, with their simple tools, as a decentralization of our big cities would be for us, with all our technology and wealth.

Further, he pointed out, urban dispersal would simply restore city dwellers to the state their ancestors had known since time immemorial: "We lived as a race for hundreds of thousands of years without cities. In fact, we have lived in cities scarcely a century; seventy-five million of us in the United States live away from cities now." While abandoning the cities might involve a "possible loss of advantage from our urban civilization"—museums, universities, orchestras, metropolitan newspapers, "aggregations of intelligentsia"—Ogburn suggested that with adequate social planning, these advantages might be equally attainable in smaller places. In fact, he wrote, the quality of life might be

> much better with well-planned smaller cities and towns. . . . We could have better health, fewer accidents, wider streets for automobiles, landing places for helicopters, more sunlight, space for gardens, more parks, less smoke, more comfortable homes, efficient places of work, and, in general, more beauty.

While serious discussion of the urban dispersal option was widespread in the late 1940s, William Ogburn's pronouncements illustrate yet again how the early discussion of the bomb's implications often moved in well- worn grooves, involving, in this case, two familiar themes in American social thought: hostility to the city, and a strong social-control impulse toward the urban masses.

CHAPTER 12

. . . The Heavens and the Earth: A Political History of the Space Age

Walter A. McDougall

Late in the sixteenth century, European states began explorations around the world that served indirectly to hasten technological improvements in a number of fields. In the nineteenth century, in the United States, an increasingly powerful state directed the spread of railroads, gas lighting, telegraph communications, and agricultural research. But not until the early twentieth century, especially not until World War I, did the state do more than promote the technology developed by private inventors or enterprises and begin to undertake the development of new technology for its own purposes. It is this state-controlled technological change, what Walter McDougall calls "command technology," that most distinguishes, he says, the twentieth century from previous centuries.

The revolutionary appearance of state-sponsored research and development (R & D) manifested itself most clearly in the ten years between 1955 and 1965, when America embarked on a program of space exploration. McDougall's book concerns these early years of the space age, when the United States began space exploration less to demonstrate humankind's "questing spirit" than to battle the Russians in the momentous struggle for international prestige.

In his book, McDougall shows that the space race began even before the Soviet launching of the Sputnik satellite in 1957, and in Sputnik's aftermath there were many different roads the United States could have taken into space. The reading selection presented here concerns a critical moment of decision, at the beginning of John F. Kennedy's administration in 1961. At this point, Project Mercury was underway, begun during Eisenhower's administration; it aimed at beating the Soviets in placing the first human in space. But in early 1961 important choices had to be made about what would be top priority beyond Mercury. Scientific research? Commercial applications? Defense? Humans on board? A lunar landing?

Interestingly, it was not the scientists and engineers who had the highest expectations of efforts to develop space technology. Instead it was lawyers, journalists, and politicians who proved to be most susceptible to what McDougall calls the "technocratic temptation," the belief that state-directed science and technology brings progressive change and can solve political problems. Elsewhere, Adlai Stevenson betrayed the prevailing power of this temptation when he said, in all seriousness, "Science and technology are making the problems of today irrelevant. . . . This is the basic miracle of modern technology. . . . It is a magic wand that gives us what we desire."

Kennedy did not immediately embrace the most ambitious plans that some proposed, to drive for a lunar landing; initially he was concerned with the possibility of adverse publicity if even the much more modest Mercury program was to meet with a setback. But the advisers the new president gathered around him to make recommendations about the U.S. space program were committed to manned spaceflight, and would persuade him to set off boldly for the new frontier in space.

Bureaucratic fighting between NASA and the U.S. Air Force influenced the course of events—to NASA's advantage. Moreover, a recommendation from the prestigious body of civilian scientists, the National Academy of Sciences, seemed to indicate that scientists had for the first time recognized the scientific value of a manned space program. The most persuasive argument for launching a lunar space program, however, was suddenly provided by the Russians: Soviet Cosmonaut Yuri Gagarin, the first human in space, orbited the earth and returned safely. (Project Mercury did not succeed in launching into space America's first astronaut, Alan Shepard, for three more weeks, and it was for a much less ambitious suborbital flight.) The domestic political furor in the United States over the Soviet's newest advance provided fresh fuel for the space race. McDougall marshals evidence to suggest that Kennedy was concerned about other foreign policy setbacks and about gaining political support for a number of domestic programs, and considered whether an expensive space program would serve as a solution to various political problems.

The final step Kennedy took before deciding on his own space policy was to appoint Vice President Lyndon Baines Johnson to prepare specific recommendations for space. Johnson's recommendations made a drive for the moon, Project Apollo, seem inescapable. But note the nature of Johnson's reasoning—international prestige and competition to secure alliances with nonaligned countries meant the space race was merely an extension of the Cold War. Reaching the moon was viewed as critical for America's global leadership, and everything depended on arriving first. James Webb, the head of NASA, and Secretary of Defense Robert McNamara, lent their weight to LBJ's recommendations for their own reasons that were just as distant from scientific criteria.

Kennedy approved the program with hardly any changes, embarking on a course whose fiscal cost remained vaguely estimated. Money was not an issue, McDougall says. Kennedy felt that whatever the cost we could not let the Soviets beat us to the moon.

It was a decision made for all the wrong reasons, McDougall tells us. But the author maintains that the decision to pursue Project Apollo entailed a hidden, and much larger cost to the country: a massive enlargement of state power. During the space race, the nation was worked up into a wartime-crisis mentality, and the increasingly desperate competition with the Soviet Union drove America toward a more centrally controlled, technocratic society. McDougall asserts that we ended up emulating the Soviet Union in order to compete with it.

Here is a glossary of abbreviations the author uses, listed alphabetically:

BoB	Bureau of the Budget
DoD	Department of Defense
FY	fiscal year
ICBM	intercontinental ballistic missile
JPL	Jet Propulsion Laboratory
NACA	National Aeronautics and Space Council
NAS	National Academy of Sciences
NASA	National Aeronautics and Space Administration
NSF	National Science Foundation
PR	public relations
PSAC	President's Science Advisory Committee
R & D	research and development
USAF	U.S. Air Force

DESTINATION MOON

"The generation that fought the war"—these were the Kennedy men. Convinced of their brilliance in comparison to the men who surrounded Ike, they extolled vigor, intellection, and movement. Behind

---

SOURCE: From . . . The Heavens and the Earth: A Political History of the Space Age by Walter A. McDougall. Copyright © 1985 by Basic Books, Inc. Reprinted by permission of the publisher.

the clichés about "company commanders" replacing the generals was the truth that World War II was the formative experience of their lives. They remembered the bitter fruits of appeasement, but above all the way war had galvanized science, industry, and government, and showed what Americans could do with technology, the proper leadership, and the inspiration of a mighty cause. Kennedy commanded a PT boat, John Kenneth Galbraith helped to draft the strategic bombing survey, Walt Rostow picked targets for armadas of B-17s and B-25s, Robert McNamara and his "whiz kids" supervised development of the huge B-29. Except for the latter, they knew little of design and production, but that was the point. Scientists and engineers, while they welcome the financial rewards flowing from political promotion of technology, are less likely to oversell it as a cure for all ills; they know their limits. Rather, it is the lawyer, economist, journalist, or politician who is most susceptible to technocratic temptation.

Or the company commander. Eisenhower had little faith in centralized management of power outside the military arena. But of these Best and Brightest, David Halberstam wrote, "if there was anything that bound the men, their followers, and their subordinates together, it was the belief that sheer intelligence and rationality could answer and solve anything." To set goals for the nation and devise methods for their achievement under state direction: this was the approach to public policy that captured university faculties and foundations in the late 1950s. Political scientists like James MacGregor Burns despaired of Eisenhower's passivity and wanted an activist presidency. FDR was the favored model. It seemed obvious that the United States could do better, that official reticence only perpetuated the ills of society, that power was not corrupting but a tool to be used for good. To set goals for the people, to assume command as the most intelligent and inspired citizens, this was simply leadership, no less.

The environment changed as well as personnel. Technological revolution was abroad in the world, and limits to action retreated beyond the horizon. In such a historical conjuncture Eisenhower's philosophy seemed not only obsolete but immoral, while a mobilized United States knew no limits. Kennedy said as much in his inaugural address: "The world is different now. For man holds in his mortal hands the power to abolish all forms of human life. . . . Let the word go forth from this time and place, to friend and foe alike, that the torch has passed to a new generation of Americans . . . we shall pay any price, bear any burden. . . ."

How different from Ike's words eight years before, when he hoped to liquidate the Korean War, slash defense spending, end regulation. Yet what expectations lay behind the eloquence? Within months Kennedy fired toward Capitol Hill a salvo of new spending measures and within a year the largest tax cut in recent memory. Only two assumptions could underpin such actions. Either a great surplus of wealth had built up in the 1950s (giving the lie to Democratic claims that the country had been "standing still" and "living off capital") or else explosive growth was expected in the coming decade sufficient to cover "any price, any burden." How could this be? Two to 3 percent growth would not yield "the revenues required for the welfare goals he had articulated, for the expanded infrastructure the cities required, or for the national security goals [Kennedy] had set," wrote Rostow. In other words, where traditional economics dictated the setting of state spending according to the ability of the economy to bear it, the new economics dictated stimulation of the economy to the point where it could sustain the desired level of spending. Kennedy and Walter Heller, chairman of the Council of Economic Advisors, forged a consensus in favor of "lifting the level of employment and the rate of growth by unbalancing the federal budget, grossly if necessary." Investment would be encouraged, wages and prices restrained by "jawboning" and a new "social contract," and business convinced that "a large, purposeful deficit" was sound policy.

Technology did not emerge from the start as a primary tool for enforced growth. But the new dogma that federal spending was beneficial to the economy and the "pay any price" mentality conditioned the Kennedy team to think of space exploration in terms of ends (were they desirable?) rather than means (can we afford it?). When a new Soviet spectacular, Third World setbacks, and the energetic advocacy of Vice President Johnson combined to force a decision on ends in space, the outcome was assured. It was Destination Moon.

For all their "space gap" talk the Kennedy men had little notion of what to do with the space program after election day. Twice in December 1960 the President-elect met with one teammate who did, and Kennedy gave to LBJ the responsibility for space in the new administration. His vehicle for doing so would be the National Aeronautics and Space Council, created by Johnson in the space act but hardly active since. Its first meeting had taken up important matters such as the transfer of JPL to NASA. But in its second meeting Eisenhower dozed off during a discussion of the NASA logo. The council met seven more times, with the President usually in attendance. But he and Glennan resolved to abolish it as early as 1959, only to be blocked by the Senate. Now Kennedy and his aide Theodore Sorensen decided to vest the chairmanship of the body in the vice presidency and provide for an executive secretariat. The post would be filled by Edward C. Welsh, economist, former aide to Symington in the 1956 air power hearings and contributor to JFK's speeches on space during the campaign.

Johnson also grasped the threads of space policy in the Senate, where he chose his successor as chairman of the Space Committee. It was Robert Kerr (D., Okla.), an oil millionaire who knew little about space but was a cagy ally. (Kerr once boasted, "I represent myself first, the state of Oklahoma second, and the people of the United States third—and don't you forget it!") With these institutional pieces in place, Johnson set out, just as in 1957, to marshal the information and influence needed to push through an accelerated space program.

What was to be done? Eisenhower had reluctantly granted the importance of prestige in space against the judgment of his scientists. Kennedy's scientific advisers also felt that prestige was overemphasized. Jerome Wiesner of MIT headed Kennedy's Ad Hoc Committee for Space and concluded, with support from the likes of Trevor Gardner and Edwin Land, that science was the only portion of the U.S. space effort free of severe defects. Their report denounced Project Mercury, which only "strengthened the popular belief that man in space is the most important aim of our non-military space effort," and held that *"a crash program aimed at placing a man into orbit at the earliest possible time cannot be justified solely on scientific or technical grounds."* The committee urged Kennedy to stop advertising Mercury lest he associate himself with a possible failure or even death of an astronaut. Instead, the U.S. government should concentrate on scientific and commercial applications such as communications satellites. The Wiesner Report comprised a scientists' critique that would echo until the moon landing and beyond.

Kennedy found the report "highly informative" and promptly named Wiesner his Special Assistant for Science and Technology—then he set the report aside. "I don't think anyone is suggesting that their views are necessarily in every case the right views." The admonition that seemed to affect the new President the most was that concerning Mercury—exploding rockets, dead astronauts, lost races—that is, not that manned spaceflight was wasteful or misguided, but that it might be a public relations failure. In his news conference on February 8, 1961, Kennedy demurred on the race for man in space, placing safety above the desire to "gain some additional prestige."

Hence the first months of the new administration showed hesitancy about space rather than bold forays into this new frontier. Kennedy was learning, Johnson preparing his ground. The PSAC opinion was already on the table, but ran counter to the visceral enthusiasm of Johnson and the Congress. The only actor missing from the scene was the new NASA administrator.

Glennan resigned in December, and Dryden, the apolitical expert, was asked to stay on as acting administrator through the transition. But what should the new man be like: a low-profile technician, a businessman, ex-general, university president, political wheeler-dealer? The choice would be a function of what the NASA chief would be asked to do. Reflecting its early confusion on this score, the transition team interviewed two dozen candidates, including James Gavin. He was an attractive choice, since he understood the R & D cycle as well as anyone, supported NASA despite his views on the military importance of the "space theater," criticized Eisenhower, and expected space technology to spark an economic revolution. But Gavin either turned down the job or was scratched as a military man. Frustrated, Kennedy tapped Johnson to fill the vacancy, Johnson consulted Kerr, and the latter touted his business partner James E. Webb—the same man who had served Truman as Director of the Budget and axed the early ICBM and satellite programs!

Webb was fifty-four years away from the rural North Carolina of his birth when he took control of the civilian space program. Trained as a lawyer in the capital, Webb became a reserve pilot in the marines and an officer in Sperry Gyroscope Company in the 1930s. He joined the Truman administration in 1946. During the Republican ascendancy, Webb had made his fortune with Kerr-McGee Oil, sat on the board of McDonnell Aviation, and given considerable time to public service. This included leading roles in the Municipal Manpower Commission devoted to urban problems, the Meridan House Foundation, a center for foreigners in the United States, and Educational Services, Inc., in which Webb collaborated on a high school physics text to meet the needs of the Space Age. In sum, he was steeped in the post-Sputnik ethic of government activism, prestige, and scientific mobilization. But when Webb arrived in Washington on a weekend late in January 1961, he told Dryden, "Hugh, I don't think this job is for me." Dryden replied, "I agree with you. I don't think it is either." Webb sent friend Frank Pace to appeal to LBJ, but he got "chased out of the office." He then saw Philip Graham, publisher of the Washington *Post:* "Phil, I've got to get out of this, can't you help me?" No, said Graham, the only man who could was Clark Clifford of the Kennedy transition team. But Clifford had also recommended Webb: "I'm not going to help you get out of it." So on Monday morning Webb reported to the Oval Office. Kennedy explained that he did not want a scientist at NASA but "someone who understands policy . . . great issues of national and international policy." Slightly mollified, Webb accepted: "I've never said no to any President who asked me to do things."

In the words of Abe Zarem, president of Electro-Optical Systems, a NASA administrator had to be

> . . . a missionary, an evangelist, with a keen sense of our national rendezvous with destiny . . . an efficient manager . . . suave, a man of exceptional social manners, particularly for briefing Congress . . . able to understand human beings to keep in effective operation people of extremely diverse personalities . . . understand the relationship between scientific knowledge and industrial might . . . know generals and admirals . . . know the "spaghetti bowl," the Pentagon, how it works and how to get around it . . . understand the workings of the Budget Bureau.

James Webb was such an extraordinary man. But there would have been no point in placing him in charge of a space program limited to small-scale science or mortgaged to the military. In fact, Webb took office on February 14 anxious "to make unmistakably clear our support for the manned spaceflight program. . . ."

The new team was in place. Lawyers, politicians, businessmen, academics, they were confident of their ability to manage a vastly expanded program of civilian command technology. But before the new team could even put the issue of NASA's future before the new President, they had to fight off another challenge to the *raison d'être* of NASA itself. The outcome of this skirmish, like the philosophy of the new administration and the choices of Johnson, Kerr, Welsh, and Webb, narrowed the possible futures of the American space program and pointed it, incredibly, toward the moon.

The melée over control of space R & D and operations after Sputnik left only two standards flying, those of NASA and the USAF. Military, and some civilian, critics still questioned the wisdom of a divided program. If military control was obnoxious, then let NASA do everything, but unify the program somehow! USAF space managers considered the verdict of 1958–59 irrational, unjust, and possibly dangerous. American rocketry grew up in the services. Farsighted officers had pleaded for years for the funds to launch the Space Age. But as soon as Sputnik had vindicated them, the government said, "OK, you were right. Now take all that you have done and hand it over to this new, civilian group." So the USAF space cadets waited, assuming the battle lost but the war still on, until the day when NASA might fade back into the status of the old NACA. This did not mean that the USAF did not cooperate; rather it must help NASA push space technology forward against the day when it might share in the spoils. In the meantime, its skillful PR apparatus advertised USAF experience and prowess in space, kept the problems of a divided program before the public, and declared that "peaceful uses of space" were best ensured by a strong U.S. military presence in orbit.

The presidential campaign, with its promise of change in the midst of "missile and space gap" mania, seemed an opportunity for the USAF to recoup. In October 1960, General Schriever established an Air Force Space Study Committee under Trevor Gardner, the man who had championed the crash program for an ICBM. Meanwhile, the Air Force Secretary's office and aerospace trade press publicized the military shortcomings of the current space program. The USAF Space Study Committee met five times over the winter and issued its top-secret Gardner Report on March 20, 1961. The first sentences revealed its position: "The military implications of the frequency and payload size of the Soviet space launches are a major cause of alarm for all members of the Committee. Under existing U.S. schedules, and with the present organization, it will be *three to five years* before we can duplicate the recent Soviet performance." Soviet men in space, orbital rendezvous, and lunar exploration posed an "impending military space threat" that could not be met by current space organization. Among the hurdles was "the insistence on classifying space activities as either 'military' or 'peaceful.'" The Soviets made no such distinctions, while American niceties only exposed the United States to political attacks. Thus the divided program was the worst of both worlds. The panel recommended that a new Air Force Systems Command be given the task of developing manned spaceflight, space weapons, reconnaissance systems, large boosters, space stations, and even a lunar landing by 1967–70. The U.S. military, after all, had a long history of leadership in exploration, and in any case the inhibitions against military spaceflight approached a unilateral arms moratorium. "The U.S.," lamented the report, "has a consistent record of underreacting to the rate of Soviet technological and military progress . . . in the military space field, we have continued to underimagine the possibilities of the future and are not yet organized to exploit them." NASA, deemed superfluous, was scarcely mentioned in the sixty-four-page report.

It probably never occurred to USAF petitioners that the new administration would embrace their ambitions, grant their military importance, and still weigh in on the side of the civilian agency. Yet the aftermath of the USAF space gambit was precisely that. Even before Gardner reported, the chairman of the House Space Committee, Overton Brooks (D., La.) sniffed the winds and preempted the assault. He told a White House conference in February that "any step-up in the [space] program must be designed to accelerate a *civilian* program of peaceful space exploration and use. . . . This is very important from the standpoint of international relations." The military had a legitimate role, but "NASA and the civilian space program badly needed a shot in the arm." A big space program would have a "pronounced and beneficial effect on America's civilian economy" as well. Three weeks later Brooks expressed to the President his serious concern about persistent rumors to the effect that radical change was about to take place in space policy in the direction of military uses.

While awaiting Kennedy's answer, Brooks sponsored hearings on DoD involvement in space. Undersecretary Roswell Gilpatric assured the committee that the DoD did not want to control NASA: "We have plenty of problems today. We don't need any more." When USAF General Thomas White took the stand, he deftly retreated. To be sure, he had spoken of NASA "combining with the military," but that was only a statement of possible fact, not of advocacy, and was meant to encourage USAF commanders to cooperate, not compete, with NASA. Chairman Brooks was pleased with this assurance but even more pleased with

President Kennedy's reply to his letter: "It is not now, nor has it ever been, my intention to subordinate the activities in space of NASA to those of the DoD."

The USAF gambit was checked: if a major expansion in space should occur, NASA would be the beneficiary. But had the USAF been frustrated as thoroughly as it appeared? Did Generals White and Schriever really hope to persuade the politicians that a single, military-run space program best met all desiderata? No one knew more about management of space R & D than Schriever, yet even as the Gardner Committee slammed the civilian program, Schriever lectured to an engineering convention in Pittsburgh on the *divergent* needs of civil and military spaceflight. They were complementary, he said, and both must be pursued with imagination and vigor. But "the military and civilian missions . . . do not merge into a single image." The technology was essentially the same, but even this was a temporary condition. First, the military would need many more space vehicles (for surveillance, communications, etc.) than NASA, since the latter's would be exploratory in nature. Second, military spacecraft would have relatively longer lives and highly repetitive missions, while scientific spacecraft changed payloads almost with every shot. Third, military vehicles must be simple, reliable, and easy to maintain, while scientific ones would be complicated. Fourth, military missions were time-critical, while NASA, certain launch windows excepted, could choose when to fund and execute projects. Finally, military space technology required close coordination between developer and user, while in NASA programs the same team of technicians served as designers and users. All this meant that different management challenges faced the two programs. Schriever predicted that NASA and USAF efforts would diverge over time, implying that a dual space program was indeed appropriate.

If these were the professional insights of the leading USAF space executive, then what are we to make of the Gardner Report he commissioned? It appears likely that the USAF deliberately overstated its case in order to educate the new administration into USAF assumptions about the Soviet military space threat. It hoped for greater, if not total, support for a "race posture," experimentation with military applications, and military participation in manned spaceflight and big boosters. General White might have been sincere in predicting large operational missions in space for the USAF, but that would come later, as the programs diverged, not at once, when R & D was still to be done. Schriever's own contribution to the policy process bears out this interpretation. He earnestly supported an accelerated NASA program, in order to stack the "building blocks" of spaceflight.

USAF background noise did make an impression. Webb's first priority at NASA was to cement ties with the DoD, while Secretary of Defense

McNamara initialed Eisenhower's planned increase in military astro-nautics for FY 1962. Abraham Hyatt of NASA then proposed a division of tasks between the two agencies that granted the DoD primacy for military missiles, reconnaissance, military communications satellites, navigation, geodesy, satellite inspection and interception, and a joint role in launch vehicles and manned spaceflight. On February 23 Mc-Namara and Webb agreed that neither agency would initiate develop-ment of new launch vehicles without the other's consent, while large solid-fueled rockets were to be a USAF show. The DoD also retained a stake in manned flight with its X-20 Dyna-Soar and was promised the opportunity to observe and learn from Mercury.

By the end of March 1961, when Kennedy finally turned to space policy, not only the Gardner Committee but also the Space Science Board of the NAS had rebutted Wiesner and come out for a vastly expanded space program. Heretofore scientific views on the space program had been hostile to "big engineering" as opposed to research satellites. But the NAS Space Science Board, chaired by Lloyd Berkner, a close friend of Webb, recommended that *"scientific exploration of the moon and planets should be clearly stated as the ultimate objective of the U.S. space program for the forseeable future."* It considered that "[f]rom a scientific standpoint, there seems little room for dissent that man's participation in the ex-ploration of the Moon and planets will be essential." The board also held (a bit beyond its competence) that "the sense of national leadership emergent from bold and imaginative U.S. space activity" pointed toward a large manned program, and that "man's exploration of the Moon and planets [is] potentially the greatest inspirational venture of this century and one in which the whole world can share; inherent here are great and fundamental philosophical and spiritual values which find a re-sponse in man's questing spirit and his intellectual self-realization."

Here was language to stoke the visionary, intellectual President! The scientifically sound but uninspiring caveats of the Wiesner Report fell flat by comparison. More important, the Space Science Board altered the terms of debate. Beforehand, the main conflict had been one of politicians and engineers pushing manned spaceflight for prestige, se-curity, or big budgets, *versus* scientists and treasurers favoring un-manned flight because of greater scientific returns and much lower costs. But now a body of scientists had come out for a manned moon program, asserted its scientific value, and appealed to something more than "knowledge gained per dollar spent." Manned spaceflight could now be viewed as something over which "good scientists disagree"; the weight of purely political judgments was accordingly enhanced.

The minutes of the Space Science Board meetings, however, tell a different story. It seems that Berkner himself proposed that the board offer an opinion on manned spaceflight, and several members who spoke in favor of it did so in hopes that it would turn nations away from weapons and war, *not* because they held it to be an efficient scientific investment. Other board members bluntly criticized manned spaceflight as misdirected. Berkner then tried to close the meeting with the observation that "this sort of negativism always appears" and that a clear-cut national decision was needed at once. Several members then disputed the wisdom of the board making any policy recommendations, but the committee finally let Berkner "pull together a statement of the Board's position." That statement was hardly representative, but Berkner assured Webb in late February that his report would support the NASA chief's recommendations on manned spaceflight.

The support was timely. Webb confronted the BoB in mid-February only to learn that "we are still pretty much in the dark as to what position the administration desires to take in the space field." But Budget Director David Bell urged Webb to do a quick review (the delay in Webb's appointment had caused NASA's budget process to slip) and recommend any changes before supplemental requests went to Congress in late March. When Webb canvassed NASA opinion, he learned that staff recommendations included acceleration of the big Saturn booster, an even bigger Nova, and the start to Project Apollo that Ike had denied. The Manned Lunar Landing Task Force under George Low even thought the NASA plan for a lunar landing "after 1970" too conservative: it could and should be done before 1970. Technical considerations were less important in selling the program, however, than political ones. "It is our responsibility," said Webb on March 17, ". . . to assess the worthwhile social objectives of our space program and to study our space effort in the context of our broad national and international goals."

The same day NASA made its first pitch to the BoB, asking for a 30 percent increase in the last Eisenhower budget. The case to accelerate, wrote Bell, "was well presented by Mr. Webb and his associates." But budget directors are professional skeptics. Seconded by Shapley, Bell questioned whether the United States should run races it might lose anyway, whether there were not better (and cheaper) ways of enhancing prestige, and whether "the total magnitude of present and projected expenditures in the space area may be way out of line with the real values of the benefits. . . ." Bell wrote this to the President, then told Dryden not to expect rapid action as Kennedy had other problems to worry about. Dryden retorted: "You may not feel he has the time, but whether he likes it or not he is going to have to consider it. Events will force this."

Webb got his first crack at the President on March 20. His presentation was a prototype of the technocratic argument he would make over and over again in years to come. He began by reminding the President of the effects of Soviet "firsts." To be sure, the Republicans had funded extensive scientific research in space (NASA's Robert Jastrow had just reported that the United States *led* the Soviets in every area of space science), but they left the United States no room for initiative, the key to which was big boosters. Furthermore, the DoD benefited from NASA programs, while in foreign policy, the civilian space program was a positive force: "We feel there is no better means to reinforce our old alliances and build new ones. . . ." But future prospects were even greater, when

> it will be possible through new technology to bring about whole new areas of international cooperation in meteorology and communications. . . . The extent to which we are leaders in space science and technology will in large measure determine the extent to which we, as a nation, pioneering on a new frontier, will be in a position to develop the emerging world forces and make it the basis for new concepts and applications in education, communications, and transportation, looking toward more viable political, social, and economic systems for nations willing to work with us in the years ahead.

Prestige, cooperation, emerging world force, viable socioeconomic systems, a new frontier—Kennedy may not have known much about space, but he knew appealing slogans. The President agreed to include $125.7 million for rockets in his defense message to Congress of March 28. Moving Saturn ahead also kept his options open; it bought time.

Two weeks later the time ran out. Yuri Gagarin orbited the earth, and American newspapers again echoed the Kremlin's judgment of it: "a psychological victory of the first magnitude"; "new evidence of Soviet superiority"; "cost the nation heavily in prestige"; "marred the political and psychological image of the country abroad"; "neutral nations may come to believe the wave of the future is Russian." In Congress, Space Committee members insisted that the administration, committed as it was to vigor, determine once and for all whether the United States was going to be first in space. James Fulton (R., Pa.) demanded public acknowledgment that "we are in a competitive race with Russia. . . ." To James Webb he said, "Tell me how much money you need and this committee will authorize all you need"; to the press corps: "I am tired of coming in second best all the time." Victor Anfuso (D., N.Y.) threatened a congressional investigation: "I want to see our country mobilized to a wartime basis, because we are at war." Chairman Brooks demanded that the White House do whatever was necessary to gain unequivocal leadership in space. Webb observed: "The committee is clearly in a runaway mood."

Kennedy's initial reaction to Gagarin was not unlike that of Eisenhower after Sputnik. His congratulatory letter to Khrushchev spoke of cooperation, and he told a press conference that "while no one is more tired than I am" of the United States being second best, he hoped "to go into other areas where we can be first and which will bring more long-range benefits to mankind." Webb, uncertain of where the President stood, lauded the broad-based scientific U.S. space program and sounded like Glennan: "The solid, onward step-by-step pace of our program is what we are more interested in than being first."

Had Kennedy and his people been genuinely contemptuous of man-ufactured prestige, they might have weathered this storm without major shifts of policy. But just two days after these initial utterances, Kennedy summoned Webb, Dryden, Wiesner, Sorensen, and Bell to the White House and invited journalist Hugh Sidey to observe as he played the leader intent on getting to the bottom of a crisis while others lost their heads. "Is there any place we can catch them? What can we do? Can we go around the moon before them? Can we put a man on the moon before them? What about Nova and Rover? When will Saturn be ready? Can we leapfrog them?" Webb assured him that NASA was moving ahead rapidly. Bell warned of the costs and Wiesner that "now is not the time to make mistakes." Kennedy assumed the burden: "When we know more, I can decide if it's worth it or not. If somebody can just tell me how to catch up. . . . There's nothing more important."

Important for what? National defense, party politics, prestige, na-tional morale? According to historian John Logsdon, Kennedy placed space within a domestic as well as foreign context. He had suffered embarrassments in Laos and the Congo, then Gagarin and the Bay of Pigs humiliation in Cuba. Somehow the trend must be reversed. But Kennedy also had a broad domestic agenda. Would an expensive space program help or hurt him in Congress? Webb believed that a big space initiative would help Kennedy with congressional power brokers and build a basis of support for all his plans. Since the space message ended up as the climax of a lengthy appeal touching on foreign and domestic programs all tied together with the Cold War ribbon, it is likely that Kennedy shared Webb's analysis.

Yet Webb himself now displayed caution. He certainly favored a big push in space, but it was he who would be responsible for it. More money for more rapid progress was one thing, but to declare a specific goal, such as those mentioned by JFK in the Oval Office, was risky. What if a moon voyage proved impossible? Or accidents should happen? This was no Lewis and Clark expedition undertaken with discretionary monies—a moon mission would mean the partial transformation of the national economy! "My own feeling," wrote Webb to Glennan on Ga-garin day, "in this and many other matters facing the country at this time is that our two major organizational concepts through which the power of the Nation had been developed—the business corporation and the government agency—are going to have to be re-examined and per-haps some new invention made." Ongoing Soviet competition required

the United States to "utilize every resource we have in education, communication, and transportation to build a more viable economic, political, and social structure for the free world. . . ." This was technocracy,
and not to be undertaken lightly. Webb was not going to get out front
on it without support from the highest quarters.

Kennedy was hardly thinking in terms of national restructuring. He
worried about prestige. But his advisers shared Webb's belief in the
growing obsolescence of free markets, balanced budgets, or limited government. Foremost among them was LBJ. Summoned to the White
House on April 19, he requested a presidential mandate to make recommendations for space. Kennedy complied:

> I would like for you as Chairman of the Space Council to be in charge of
> making an overall survey on where we stand in space. . . . Do we have a
> chance of beating the Soviets by putting a laboratory into space, or by a
> trip around the moon, or by a rocket to land on the moon, or by a rocket
> to go to the moon and back with a man? Is there any other space program
> which promises dramatic results in which we could win . . .?

The next day Kennedy announced the study to the press. The hinge
was not cost—he admitted that "billions" were involved—but "whether
there is any program now, regardless of its cost, which offers us hope
of being pioneers in a 'project." When asked if the United States should
beat the Russians to the moon, he replied, "If we can get to the moon
before the Russians, we should."

Johnson now had carte blanche to set a goal. Rarely had a great
political issue been so clear cut, but rarely had the variables been so
obscure: Could giant rockets be built? Could men stand long periods of
weightlessness? Could orbital rendezvous be mastered? Was the lunar
surface suitable for soft landing? What were the Soviets up to? Johnson
consulted NASA, which saw a chance of beating the Soviets to manned
circumlunar flight or a lunar landing, but at the cost of at least $11.4
billion extra dollars over ten years. Then he asked McNamara, three
business cronies, von Braun, Schriever, and Vice Admiral John T. Hay-
ward. They all supported a moon landing, the last two stipulating only
that it not detract from military missions. Von Braun even spoke of
putting all other elements in the space program "on the back burner."
One of the businessmen, Donald Cook, believed that an action must be
"based on the fundamental premise that achievements in space are
equated by other nations in the world with technical proficiency and
industrial strength . . . and will be of fundamental importance as to
which group, the East or the West, they will cast their lot. . . ."

Johnson sent the President a report so loaded down with assumptions that a moon landing was the inescapable conclusion: (1) the Soviets
led the United States in prestige; (2) the United States had failed to

marshal its superior technical resources; (3) the United States should recognize that countries tend to line up with the country they believe to be the leader; (4) if the United States did not act, the Soviet "margin of control" would get beyond our ability to catch up; (5) even in areas where the Soviets led, the United States had to make aggressive efforts; (6) manned exploration of the moon was of great propaganda value but was essential whether or not the United States was first. In another context, LBJ put it more pithily: "One can predict with confidence that failure to master space means being second-best in the crucial arena of our Cold War world. In the eyes of the world, first in space means first period; second in space is second in everything."

The persuasive vice president then worked on Webb in a "consultative meeting" packed with lunar zealots: Dryden, Welsh, the three businessmen (including Frank Stanton of CBS), Senators Kerr and Bridges. All sensed that the moment for NASA had arrived. Why did Webb, of all people, hang back? He said he wanted to be sure that NASA had enough support and really knew what it was getting into: if NASA appeared to be the initiator of an expensive, questionable project, it might be left twisting in the wind should the national mood change or delays and failures ensue. NASA must be *given* the task, asked to do this risky thing by an anxious nation. Then Webb would have the leverage, down the road, to claim the backing needed to see the agency through harder times. (Indeed, five years later, Webb wrote to then-President Johnson: "You will remember that in the sessions you had in 1961 with your advisers and Congressional leaders, I was quite reluctant to undertake the responsibility of building a transportation system to the moon and that you had to almost drive me to make the recommendation which you sent on to President Kennedy." For at that May 3, 1961, meeting, LBJ was "close to demanding that NASA recommend for Apollo.")

The next day Webb wrote LBJ that he was ready to climb on board. "I think I can say also that my main effort yesterday was to be certain you and the Senators were under no illusions whatever as to the magnitude of the problems involved in carrying out this decision and the absolute necessity, in my opinion, for a decision to back Secretary McNamara and myself to the limit. . . ." Congress and the press would ride NASA "like two packs of hounds." He must be sure of the President and the Vice-President. Thrice he insisted, and thrice he referred to McNamara.

Why McNamara? Why did the Secretary of Defense endorse a NASA moon program? He knew the importance of prestige and of a "peaceful, civilian" space effort. But the value of a moon mission was not self-evident. McNamara's internal studies in fact convinced him that some of the more "way out" space programs were ripe for the axe, and as late as March 23 he expressed preference for a "normal rate of investigation" and "stated rather emphatically that he would accord a higher priority to items included, about to be excluded, and already excluded from the Defense budget than he would to the programs in question." But there were still larger considerations, including the very health of the aerospace industry. McNamara knew that the missile gap was dubious and believed that defense R & D was out of control and that the aerospace lobby was the main obstacle to his plans for modern cost-accounting. But the sort of cutbacks he envisioned might damage an industry that had expanded tremendously in the 1950s on the strength of jet aviation and the space boom. What was more, government financing of idle plant capacity, cost-plus contracting, and "progress payments" (or "get paid as you go") to contractors were all being phased out. These measures hit the industry just as new plant, equipment, and personnel were needed to participate in the space revolution. The result was a profit squeeze despite record sales and a *tenfold* increase in aerospace debt in the decade after 1957. Aviation was never an easy business in which to make money, but in the early Space Age the industry superimposed high financial risk upon high business risk, a classic violation of sound corporate finance.

The health of the aerospace industry was as much a government worry in 1961 as in 1947. Fairchild, General Dynamics, mighty Lockheed, and Douglas were already in trouble, the latter having lost over $100 million in 1959–60. McNamara's pet reforms, including fixed-price contracting, cost-benefit analysis even on R & D programs, and insistence on high definition even of early design work all promised to trim further the narrow margin the industry trod. In this context an expensive space program, rich in new technology but lying outside the reach of the USAF and his own budget, must have seemed to McNamara, on second thought, felicitous. Giving Apollo to NASA would please the aerospace lobby and Congress, while the USAF, bereft of its allies, would nurse its jealousy alone.

Now everyone was on board. The Saturday after Alan Shepard's flight found NASA, DoD, and BoB delegations gathered to discuss the least mundane of political topics: going to the moon. In the morning session Webb and McNamara exchanged reports done at LBJ's behest. In the afternoon they tackled Apollo. McNamara approved it. He believed the USAF was "out of control": Apollo would help relieve military, industrial, and congressional pressure on him so that he could get on with the tough decisions on defense programs. He also believed, and everyone concurred, that large space projects "reflect the capacity and will of the nation to harness its technological, economic, and managerial resources for a common goal." Apollo was proposed; no one dissented.

The secret "Webb-McNamara Report" originally contained a preamble, drafted by John Rubel of the DoD, adumbrating McNamara's references to the USAF and aerospace industry. It was cut in favor of the habitual rhetoric justifying spaceflight on the basis of science, commerce, defense, and especially prestige. Soviet attainments, the report suggested, were the result of a program planned and executed at the national level over a long period of time, while the United States had "over-encouraged the development of entrepreneurs and the development of new enterprises." The United States must kick its tendency to embellish its designs. "We must insist from the top down that, as the Russians say, 'the better is the enemy of the good.'" Buried in the report was this conjecture: "It is possible, of course, that the Soviet program is not actually the result of careful planning toward long-range goals. . . . Perhaps luck played an important part. . . ." But the evidence pointed "dramatically" in the other direction. "Of all the programs planned, perhaps the greatest unsurpassed prestige will accrue to the nation which first sends man to the moon and returns him safely to earth."

Kennedy first saw the report on May 8 and met with Cabinet members on the tenth. The BoB still worried about setting dates for a moon landing and projects aimed at prestige rather than technological advance per se, not to mention the costs. Economic advisers and Secretary of Labor Arthur Goldberg even denied that the space program would stimulate the economy. But, as McGeorge Bundy recalled, "the President had pretty much made up his mind to go," and, as Wiesner recalled, when McNamara showed the President that *without* Apollo a definite oversupply of manpower would exist in the aerospace industry, "this took away all argument against the space program."

Johnson returned from a trip to Southeast Asia on May 24. A letter from Webb awaited him: "The President has approved the program you submitted, with very few changes, and the message will go up on Wednesday."

That message asked Congress to spend upward from $20 billion on command technology for a political goal. Compared with 1946, when the Atomic Energy Act was assailed as totalitarian, or 1948, when funding for the NSF was restricted by law, or even 1958, when Ike shied from racing in space and struggled to restrict the terms of the education act, Apollo signaled a new age. The technology race that began with weaponry now extended to a civilian pursuit, held in turn to be a symbol of overall national prowess. Where the Eisenhower men doubled and tripled spending on science, education, and R & D, it was their intention to contain as far as possible the effects on traditional values and social institutions and the relationship of the public and private sectors. The men who launched Apollo came to office dissatisfied with existing state management of the national treasure and talent, and began to view the space program as a catalyst for technological revolution, social progress, and even the "restructuring of institutions" in ways that were dimly foreseen but assumed to be "progressive."

How this change occurred in so short a time is not a mystery, but rather that most vexing of historical problems, the "overdetermined event." New men arrived and brought with them those ideas of the "seed time" of the 1950s. Among those ideas were the notions that the Third World was the main theater of the Cold War and that in that contest prestige was as important as power. Their new ideas validated a far greater role for government in planning and executing social change. The new men also cared more for imagery and felt increasing pressure to display their control over affairs in the wake of early setbacks in foreign policy. Finally, each major figure in space policy—Kennedy, Johnson, Webb, Dryden, McNamara, Welsh, Kerr, and others—saw ways in which an accelerated space program could help them solve problems in their own shop or serve their own interests. This is not to say that they were petty; it is to say that they were technocratic, applying command technology to political problems.

As for contrary arguments, they were disposed of, one by one. Nixon himself abandoned the original Republican skepticism toward space races even before the campaign. The USAF view that the main threat was military got nowhere with the image-conscious civilians. The scientists' argument against prestige-oriented manned spaceflight was bulldozed. When the Soviets weighed in by orbiting Gagarin, and the Shepard flight confirmed NASA's contention that the mission was feasible, all barriers came down. All, that is, except cost, and that, too, was less important in the new White House. We will probably never know precisely what was in Kennedy's mind when he decided that Americans should go to the moon. What may have tipped the balance for him and for many was the spinal chill attending the thought of leaving the moon to the Soviets. Perhaps Apollo could not be justified, but, by God, we could not *not* do it.

Of all those who contributed to the moon decision, the ones farthest in the background were the engineers of Langley and Goddard and Marshall, many of whom devoted their lives to spaceflight, designing dreams. Their reports and studies were necessary buttresses to the political arguments: they had to persuade that the thing could be done. Otherwise, they were absent. Some of their visionary talk about exploration and destiny found place in political speeches, but their efforts to stretch the minds and hearts of their fellows, to sow wonder for its own sake, got lost in their very adoption by the technocratic state. What Constantine's conversion did to the Christian church, Apollo did to spaceflight: it linked it to Caesar. The new faith might conquer the empire, but its immaculate ability to stir hearts was accordingly diminished. Of course, it could not have been otherwise.

CHAPTER 13

More Work for Mother: The Ironies of Household Technology from the Open Hearth to the Microwave

Ruth Schwartz Cowan

It is sobering to note the recentness of the arrival of many household appliances and amenities in most homes in the United States. As recently as 1940, one out of every three Americans was still carrying water in buckets, and two out of three Americans did not enjoy the comforts of central heating. In 1941, only 52 percent of American families owned or had "interior access" to a washing machine, thirty years after the machines had first come on the market and twenty years after their prices had fallen as the result of mass production. About the same percentage of families had mechanical refrigerators (the successor to the older, literal "iceboxes.") And as many as one third of all households were still cooking with wood and coal, which meant backbreaking labor to provide the fuel and attend to the requisite cleaning.

By the early 1980's, running water and built-in bathrooms, central heating, washing machines, refrigerators, and gas or electric stoves had become basic household items. It would seem that a dramatic technological revolution of sorts had taken place close to home. Yet Ruth Schwartz Cowan argues that all of this added up to little diminishment of the housework burden most women shouldered.

Cowan offers several interrelated explanations of this apparent paradox, in this selection, taken from the concluding chapter of her study of American housework. She points out that the technological revolution hastened the disappearance of two important sources of assistance that even working-class housewives formerly had been able to afford: paid domestic labor and commercial delivery services. Maids and home-delivered groceries became faint memories for most. Consequently, housewives of all economic classes discovered that they faced housework unassisted. In the early twentieth century, a college professor's wife, for example, had two household assistants, but no assistants are to be seen in the 1950s, when we hear a latter-day college professor's wife narrate her typical day.

The new household technologies made work at home much less physically onerous, but surprisingly they did not significantly reduce the amount of time American women spent doing housework. Cowan argues further that wives remain, for the most part, unassisted by their husbands, who, if anything, shoulder even less of the housework burden than before because of the arrival of new appliances. Despite the journalistic hoopla in the 1950s and late 1970s about "new husbands" and "househusbandry," Cowan remains unconvinced that the gender division of work actually has become more equitable in most households.

Cowan describes the reasons why women decided to enter the paid labor force—some by choice, others by necessity. But all "working" women, she maintains, faced stigmatization in a society that still clung to the notion that a woman's place remained in the home. Theoretically, "functionalist" sociologists might have been expected to provide some encouragement to women who got outside jobs. These sociologists believed that economic and technological changes had rendered the household "functionally" useless, but instead of urging women to move out of the "obsolete" household, they instead urged women to concentrate on the remaining "feminine" functions, such as nurturing the children. A "backward search for femininity" led couples to decide to have many children. And some housewives took up preindustrial crafts and skills in an attempt to return to an idealized past, before modern appliances challenged a housewife's traditional "function."

Given the strong negative connotations of "working mother," women with children who entered the labor force had to fight tremendous ideological pressure to stay at home. A technological explanation of why these women and the others nevertheless took up outside work is that the revolution in household technology left housewives with free time. But Cowan argues that this explanation does not stand. It was the housewives who could least afford the new labor-saving devices and amenities who were the first to enter the labor market. Cowan prefers to treat the new conveniences and appliances as mere "catalysts" of women's participation in the work force, not causes.

In the concluding section, the author reviews the place of female housework in American history. It has always been, and continues to be, disregarded by economists because it is hard to measure. It has always been, and continues to be, defined as the proper domain of women. Cowan makes a provocative argument about the postwar growth of our suburbs and the technological systems that make them work: she says that the modern individual home is outfitted and located in ways that make it very difficult *not* to have someone on hand to ferry the children to and fro and take care of all the still-considerable amount of housework that is socially defined as necessary. And when this certain "someone" decided to take a new full-time job outside the home, she discovered that she was still stuck with the old one at home. Hence, the classic double burden carried by working women.

The technological systems and low-density housing patterns that are in place are here to stay, so a certain amount of domestic work seems destined to remain. But Cowan prescribes elimination of what she regards to be "makework" domestic duties, which originate from old cultural standards that were established to announce the family's distance from poverty or to keep the separate domains of women and men clearly distinct, free from "sexual pollution." In short, Cowan says, the technology itself is not to be faulted; rather we should use it in ways that at long last make housework less time-consuming, and less the special burden of women.

HOMOGENIZING HOUSEWORK

Over, under, around, and through those statistics about the technological systems with which we live, lies a daily reality about the work processes of housework that we often forget. If the basic material conditions of life have become homogenized for all Americans (the fact that the less-than-basic material conditions have not is another matter, relevant to another book), so have the work processes of housework. In times past, housewives of the "uncomfortable" classes were manual laborers in their own homes, but housewives of the "comfortable" classes were both managers and laborers. Nowadays, the general expansion of both the economy and the welfare system has led fewer people than ever before into the market for paid domestic labor, and the diffusion of appliances into households, and of households into suburbs, has encouraged the disappearance of various commercial services. The end result is that housewives, even of the most comfortable classes (in our generally now comfortable population) are doing their housework themselves. Similarly, the extension of schooling for those who are young,

---

SOURCE: From *More Work for Mother: The Ironies of Household Technology from the Open Hearth to the Microwave* by Ruth Schwartz Cowan. Copyright © 1983 by Basic Books, Inc. Reprinted by permission of the publisher.

the proliferation of school-related activities, and the availability of jobs for those who have finished their schooling has led to the disappearance of even those helpers upon whom the poverty-stricken housewife had once been able to depend. Hence, in almost all economic sectors of the population (except the very, very rich), housework has become manual labor: the wife of the lawyer is just as likely to be down on her hands and knees cleaning her kitchen floor as the wife of the bricklayer or the garbageman. In 1914, the wife of a college professor had, as I described in chapter 6, two different kinds of household assistant (a laundress, who washed and did heavy cleaning; a student who cleared after meals, did light cleaning, and supervised the children when their mother was away) and did much of her marketing over the telephone. Forty years later, the wife of another college professor described her typical day this way:

> I get up at 6 A.M. and put up coffee and cereal for breakfast and go down to the basement to put clothes into the washing machine. When I come up I dress Teddy (1-½) and put him in his chair. Then I dress Jim (3-½) and serve breakfast to him and to my husband and feed Teddy.
>
> While my husband looks after the children I go down to get the clothes out of the machine and hang them on the line. Then I come up and have my own breakfast after my husband leaves. From then on the day is as follows: Breakfast dishes, clean up kitchen. Make beds, clean the apartment. Wipe up bathroom and kitchen floor. Get lunch vegetable ready and put potatoes on to bake for lunch. Dress both children in outdoor clothes. Do my food shopping and stay out with children until 12. Return and undress children, wash them up for lunch, prepare lunch, feed Teddy and put him to nap. Make own lunch, wash dishes, straighten up kitchen. Put Jim to rest. Between 1 and 2:30, depending on the day of the week, ironing (I do my husband's shirts home and, of course, all the children's and my own clothes), thorough cleaning of one room, weekend cooking and baking, etc.; 3 P.M., give children juice or milk, put outdoor clothes on. Out to park; 4:30 back. Give children their baths. Prepare their supper. Husband usually home to play with them a little after supper and help put them to bed. Make dinner for husband and myself. After dinner, dishes and cleaning up.
>
> After 8 P.M. often more ironing, especially on the days when I cleaned in the afternoon. There is mending to be done; 9 P.M., fall asleep in the living room over a newspaper or listening to the sound of the radio; 10 P.M., have a snack of something with my husband and go to bed.

And just as striking were the comments of another housewife in the same decade—a twenty-four-year-old woman living in the then newly build Levittown, Pennsylvania; she was described by those who interviewed her as a member of the "working class." This housewife, whose grandmother might well have been grateful to have bread and soup on the table at night, described her day in terms virtually identical to those of the college professor's wife:

> Well, naturally, I get up first, make breakfast for my husband and put a load of clothes in my washer while breakfast cooks. Then I wake him up, give him his breakfast and he's off to work. Then I make breakfast for the children. After the children eat I dress them and they go out to play. Then I hang the clothes up and clean lightly through the house. In between times I do the dishes—that's understood of course. Then I make lunch for the children and myself and I bring them in, clean them up, and they eat. I send them out to play when they're done and I do the dishes, bring the clothes in and iron them. When I'm done ironing it's usually time to make supper, or at least start preparing it. Sometimes I have time to watch a TV story for half an hour or so. Then my husband comes home and we have our meals. Then I do the dishes again. Then he goes out to work again—he has a part time job—at his uncle's beverage company. Well, he does that two or three nights a week. If he stays home he watches TV and in the meantime I get the kids ready for bed. He and I have a light snack, watch TV a while and then go to bed.

In the 1950s (and the 1980s) the housewife of the "professional classes" and the housewife of the "working classes" were assisted only by machines. Few such women had paid household help, and fewer still had food or milk or clean laundry delivered to their doors. The differences between these women were no doubt profound—differences in levels of education, in families of origin, in annual household income; but those profound differences did not produce, as they would have done in the past, equally profound variations in the ways in which the women did their work.

Apparently, also, there were no significant variations in the time that women spent at that work. One sophisticated statistical analysis of time-use data collected from a large national sample of households in 1965 found that the average American woman spent about four hours a day doing housework (or twenty-eight hours a week) and about three and one-half hours a day (or twenty-six and a half hours per week) caring for children (a fifty-four-hour week). These figures were startling in two respects. First, they were not strikingly different from what Leeds had found for affluent housewives in 1912 or from what other researchers had reported for rural and urban housewives in 1935. Second, these averages were not markedly affected either by the income level of the household or by the educational attainment of the housewife: women who managed on less than four thousand dollars a year in household income spent 245 minutes per day at housework and 207 at child care; while, at the other end of the income scale, housewives who could dispose of over fifteen thousand dollars put in 260 and 196 minutes at

housework and child care, respectively. Housewives with college edu-
cations were logging in 474 minutes a day of housework and child care
(a little under eight hours); and housewives who had not completed
grade school put in almost equally tiring days of 453 minutes (or seven
and one half hours).

Neither the working-class wife nor her middle-class contemporary
could have expected her husband to help much with this work. For a
while, in the 1940s, there was a hullabaloo in the popular press about
"new husbands" in suburbia who were diapering babies and drying
dishes and cooking barbecues and otherwise becoming "feminized."
Again, in the late 1970s, a spate of books and national magazine articles
appeared touting the virtues of "househusbandry," most of these articles
written, it turned out, by free-lance writers and journalists who had
decided to stay home for a while with their children when their wives
went back to work. If the results of sociological studies are to be trusted,
not much lay behind either one of those journalistic episodes. Men do
very little housework; and the few "househusbands" there have ever
been seem not to have stuck to it for long. Whether men are asked to
estimate the time that they spend at housework, or wives are asked to
estimate their husbands' time, or outside observers actually clock the
amount of time that men spend at it, no one has ever estimated men's
share of housework at anything higher than one and a half hours per
day. Housewives who are not employed in the labor market spend,
roughly speaking, fifty hours a week doing housework; housewives who
are employed outside their homes spend, again roughly speaking, thirty-
five hours on their work in and for their homes. Men whose wives are
employed spend about ten minutes more a day on housework than men
whose wives "stay home", and men who have small children add yet
another ten—a grand total, for these particularly helpful husbands, of
just under eleven hours of housework a week. Men who do housework
tend not to do the same work that their wives are doing: they take out
the garbage, they mow the lawns, they play with children, they occasionally go to the supermarket or shop for household durables, they
paint the attic or fix the faucet; but by and large, they do not launder,
clean, or cook, nor do they feed, clothe, bathe, or transport children.
These latter—the most time-consuming activities around the home—
are exclusively the domain of women. In households that are particularly
well equipped with appliances, men do even less housework, partly
because they believe that the work simply cannot be onerous, but also
because some of the "extra" appliances actually relieve them of sex-
related, or sex-acceptable chores. In homes where there are garbage
disposals, men give up removing the small quantities of garbage that
still need to be carried to the curb; and in households where there are

dishwashers, men cease providing whatever help with the dishes they had formerly proffered.

Thus, there is more work for a mother to do in a modern home because there is no one left to help her with it. Almost all of the work that once stereotypically fell to men has been mechanized. Families tend to live a considerable distance from the place where the male head of the household is employed; hence, men leave home early in the morning and return, frequently exhausted, late at night. Children spend long hours in school and, when school is over, have "after-school activities," which someone must supervise and from which they must be transported. Older children move away from home as soon as they reasonably can, going off to college or to work. No one delivers anything (except bills and advertisements) to the door any longer, or at least not at prices that most people can afford; and domestic workers now earn salaries that have priced them out of the reach of all but the most affluent households. The advent of washing machines and dishwashers has eliminated the chores that men and children used to do as well as the accessory workers who once were willing and able to assist with the work. The end result is that, although the work is more productive (more services are performed, and more goods are produced, for every hour of work) and less laborious than it used to be, for most housewives it is just as time consuming and just as demanding.

THE "WORKING" MOTHER

The modern technological systems on which our households and our standard of living depend were constructed on the assumption that women would remain at home, that they would continue to function as pre-industrial workers (without paychecks, time clocks, or supervisors), and that, as a corollary, they would not be tempted to enter the labor market except under unusual (and usually temporary) circumstances. Ironically, the last of these assumptions proved erroneous. In the postwar years, more and more married women, and more and more mothers, entered the labor force, the comforts of full-time wifehood and motherhood and the existence of washing machines and dishwashers notwithstanding.

In the decades after the Second World War, the national economy shifted its focus from production to service, from manufacturing to communication; and, in the process, jobs were opened up for which women were considered to be appropriate candidates: jobs as typists, clerks, and receptionists; as waitresses, store clerks, and stenographers; as teachers, social workers, nurses, administrative assistants; and, later, as computer programmers. To various women, at various times, those jobs and the salaries they provided, proved to be attractions too great to resist. In different households, the decision that wife and/or mother would "go back to work" or "continue working" was made at different times, determined either by what was going on in the world outside the family or by a particular family's development. Some women "continued to work" in the postwar years because they were reluctant to give up the life and the income to which they had become accustomed during the war; some women went home and had babies and did not re-enter the labor force until their children were grown and out of the house; other women never went back to work. As the years passed, some younger women decided not to interrupt their careers when their babies arrived, because the high level of education that they had attained, and the high salaries that they could consequently hope to command, seemed to compensate for the double burden of motherhood and career which they had to shoulder. Other women found that, whether or not they were graced with higher education and higher incomes, the growing pressure of inflation was so seriously eroding the purchasing power of their husbands' income that, small children or no, they had to go back to work. Furthermore, as a result of divorce, desertion, or the decision to remain single, other women, in increasing numbers, had no husband's income to fall back upon. The end result was that, by 1980, just over 40 percent of the total workforce was female (up from 25 percent in 1940), women with children at home constituted almost 20 percent of the labor force, and more than half of the nation's children under the age of six had mothers who were working full time. Even though different women achieved the status of being "homemakers with jobs" at different times, very large numbers of them did achieve it; and if present trends continue unabated, even more of them will do so in the future.

Woman's "Place"

It is hardly surprising that, in the immediate postwar years, many women struggled mightily with the decision to take a job, since cultural pressures of the most extraordinary kind were being brought to bear against the employment of wives and mothers. If many husbands and children opposed that decision even before they had had a chance to discover its consequences, they, too, can barely be blamed, since the public debates on the subject gave them not the slightest reason to believe that the venture would end successfully. In the 1950s and the 1960s, psychiatrists, psychologists, and popular writers inveighed against women who wished to pursue a career, and even against women who wished to have a job, and referred to such "unlovely women" as "lost," "suffering from penis envy," "ridden with guilt complexes," or just plain "man-hating." Mass-circulation magazines almost never depicted a working wife, unless to paint her in derogatory terms: working mothers were blamed for the rise in juvenile delinquency in the 1950s, for the soaring divorce rate of the 1960s, and for the rise in male impotence in the 1970s. Women's magazine fiction of the day was populated by "glowing" pregnant women and "barren" working women, whose "hungers were not yet appeased, whose destinies were not yet fulfilled"; by children who felt abandoned when their mothers were not there to greet them on the day the teacher had finally given them an "A"; and by husbands who, while tempted by the career women in their offices, always returned to their less glamorous, but more feminine wives with a warm smile and a rose behind their backs. Betty Friedan, who worked for and wrote for some of those magazines in the postwar years, recalls:

> When you wrote about an actress for a woman's magazine you wrote about her as a housewife. You never showed her doing or enjoying her work as an actress, unless she eventually paid for it by losing her husband or her child, or otherwise admitting failure as a woman.

Friedan might well have added that newspaper and magazine profits depended upon the sale of advertising space to manufacturers and retailers of consumer goods; and that in the postwar years, many advertising specialists and market researchers, who advised the manufacturers and the retailers, viewed the working woman as someone who was either too poor or too preoccupied to spend time and money in the stores. Hence, profit-conscious editors, and the writers who desired their custom, were not inclined to enhance the image of the working wife, even if they happened to be one themselves.

Sociologists and other academic social scientists, rather than be left on the sidelines, joined in the debate about women's proper place by adopting what has come to be called the "functionalist" interpretation of the recent history of the family and then by broadcasting that interpretation in countless textbooks and lectures. As I have explained in chapter 4, this argument suggested that since industrialization began, households have been deprived of their essential productive roles in the economy and, consequently, housewives have been deprived of their essential productive functions. Modern women are in trouble, the analysis continued, because modern technology has either eliminated or eased most of their earlier burdens, but modern ideologies have not kept pace with the change. One solution to the problem, the social scientists noted, would be for women to take their place in the market economy; but this solution, many of the experts argued, would be contrary to female instincts and biological needs and would interfere with the few remaining functions that housewives still perform at home—namely, socialization of young children and tension management. A better solution would be to create a new ideology, one that would rationalize the woman's situation and diminish the likelihood that she would suffer "role anxiety."

The "Backward Search for Femininity"

Ironically, the ideology that became popular in the years when functionalism dominated sociology constituted a symbolic (but only a symbolic) reflection of the very set of conditions that had made it possible for many Americans to have the comfort both of indulging in ideological pursuits and of attending lectures in sociology. One perceptive observer referred to this ideology as the "backward search for femininity." If women who lived before the Industrial Revolution had led happy, fruitful, and productive lives (as the sociologists were suggesting), then it seemed reasonable to assume that modern discontents could be wiped away if women would return at least to some of the conditions that had pertained in Martha Washington's day. In communities across the land (especially in those that were particularly affluent and, therefore, farthest removed from the horrors of pre-industrial conditions), people were acting out the sociologists' prescriptions by bearing numerous children (the baby boom appears to have been a result of a deliberate decision on the part of affluent couples to have more children then their parents had), by breastfeeding those numerous children, raising vegetables in their backyards, crocheting afghans, knitting argyle socks, entertaining at barbecues, hiding appliances behind artificial wood paneling, giving homemade breads for Christmas presents, and decorating their living rooms with spinning wheels. "I interviewed a woman," Betty Friedan reported,

> in the huge kitchen of a house that she had helped build herself. She was busily kneading the dough for her famous homemade bread; a dress she was making for a daughter was half-finished on the sewing machine; a handloom stood in one corner. Children's art materials and toys were strewn all over the floor of the house, from front door to stove: in this expensive modern house, like many of the open plan houses in this era, there was no door at all between kitchen and living room. Nor did this mother have any dream or wish or thought or frustration of her own to separate her from her children. She was pregnant with her seventh; her happiness was complete, she said, spending her days with her children.

The wiles of the "backward search for femininity" apparently enticed men as well as women—as is nowhere more strikingly illustrated than in the writings of Kurt Vonnegut, whose novels ruthlessly dissect postwar mentality. In *Player Piano* (1952), Vonnegut created Paul Proteus, an archetypically unhappy "organization man" (an engineer working for a big electrical manufacturing company), who lives with his wife,

Anita, in an archetypically "backward looking" home, replete with a huge fieldstone fireplace with candle molds over the mantel:

> Paul narrowed his eyes, excluding everything from his field of vision but the colonial tableau, and imagined that he and Anita had pushed this far into the upstate wilderness, with the nearest neighbor twenty-eight miles away. She was making soap, candles, and thick wool clothes for a hard winter ahead, and he, if they weren't to starve, had to mold bullets and go shoot a bear. Concentrating hard on the illusion, Paul was able to muster a feeling of positive gratitude for Anita's presence, to thank God for a woman at his side to help with the petrifying amount of work involved in merely surviving. As, in his imagination, he brought home a bear to Anita, and she cleaned it and salted it away, he felt a tremendous lift—the two of them winning by sinew and guts a mountain of strong, red meat from an inhospitable world. And he would mold more bullets, and she would make more candles and soap from the bear fat, until late at night, when Paul and Anita would tumble down together on a bundle of straw in the corner, dog-tired and sweaty, make love, and sleep hard until the brittle-cold dawn.

Such erotic and historical fantasies were (and still are) potent cultural forces; they help us to understand not only why some people have difficulty coming to terms with the reality of their lives, but also why some people (most notably affluent housewives) are still spending so very much time at their work. People who believe that family solidarity can be bolstered by hand-dipped chocolates and hand-grown string beans are bound to spend a lot of time dipping chocolates and growing string beans.

In any event, even if these ideological props for full-time house-wifery had not existed, historical experience itself would have militated against widespread enthusiasm for the entry of married women into the labor force. The adults who were worrying about these matters in 1950 (and even in 1960) had been children of the Depression; hence, they had good reason to remember that in their youth a "working mother" had been a person to be pitied, and her family had quite possibly been a family to be shunned. If "mother worked" during the 1920s and the 1930s, her family was more than likely to be poor, the father more than likely to be unemployed, the children more than likely to be dirty, the house more than likely to be in disrepair; when "mother worked," there were children who had no one to nurse them through illnesses, meals that were hastily thrown together from whatever could be found ready-made in the markets, poor teeth, clothing that did not fit, dirty floors, skin rashes, and bad breath. It hardly mattered that only a few of these symptoms of poverty were likely to have been directly attributable to the mother's employment, because the fact of her employment served as a symbol for all of them. Similarly, at the other end of the economic scale, the presence of a full-time housewife served as symbol not just for the status of the family, but also for its degree of good health and for its decent living standards. Whether she actually did the work or whether she directed the work that was to be done, the presence of a full-time wife and mother meant careful supervision of the family's health, a well-appointed living room, white stockings, ironed hair ribbons, regular church attendance, Sunday dinner, birthday parties. All those small (and large) comforts both helped to demonstrate the family's status and to ensure that it did not fall. The postwar working-class husband who complained that he would be embarrassed in front of his friends if his wife went out to work, was as much a product of this historical experience as his middle-class contemporary who claimed that two well-organized dinner parties a month would do more for his family's annual income than the salary his wife would be able to earn at a job.

The Role of the Machine

In the end, whatever the complaints of husbands may have been (and there were many of them), and however ambivalent wives and mothers may have felt (as many of them did), by the time the children of the baby boom had come to maturity, the "working mother" had become the "normal American housewife"; and many people believed that the widespread diffusion of modern technology was, in and of itself, responsible for this transformation. On common-sense grounds alone, a causal connection between the washing machine and the working wife seems justified: if it takes less time to do the wash with a Bendix than it did with a washtub, and to cook a meal since the advent of Birdseye, then housework must take less time (and certainly less energy) than it used to, and women must thus be tempted to fill their free time with paid employment.

The only trouble with this argument is that one empirical investigation after another has failed to find evidence for it; common sense, in this case as in many others, is not a reliable guide to the truth. As we have seen, even with washing machines and frozen vegetables, housewives do not have much free time; 50 hours per week is ten hours more than what is now considered the standard industrial week. Housewives began to enter the labor market many years before modern household technologies were widely diffused; and the housewives then entering the workforce were precisely those who could not afford to take advantage of the amenities that then existed. Even in the postwar labor market the sociological variable that correlates most strongly with a married woman's participation in the labor force is her husband's income. And the correlation is strongly negative: the housewives who are most likely to enter the labor market are the ones who are least likely to have many labor-saving devices and household amenities. Indeed, in the early post-war years, some married women were entering the labor force precisely in order to acquire those attributes of affluence.

Where the sociologists and economists have failed to find a causal connection, the historians may be able to suggest a substitute. The washing machine, the dishwasher, and the frozen meal have not been *causes* of married women's participation in the workforce, but they have been *catalysts* of this participation: they have acted, in the same way that chemical catalysts do, to break certain bonds that might otherwise have impeded the process. Most American housewives did not enter the job market because they had an enormous amount of free time on their hands (although this may have been true in a few cases). Rather, American housewives discovered that, for one reason or another, they needed full-time employment; and subsequently, they discovered that, with the help of a dishwasher, a washing machine, and an occasional frozen dinner, they could undertake that employment without endangering their family's living standards. The symbolic connection between "working wife" and "threatened family" was thus severed, not by ideologues but by housewives with machines. Working mothers discovered that, although they were weary when they left the office or factory, they could still manage to get a decent dinner on the table that night and clean clothes on everyone's back the next morning. Husbands discovered that they had been deprived of few, if any, of the comforts to which they had become accustomed, and that additional comforts (namely, ones connected with having more cash on hand) had appeared. Children discovered that they could, if need be, make their lunches and their breakfasts themselves.

Viewed from a national perspective, American housewives entered the labor market without destroying either the level of health or the level of comfort to which they and their parents had become accustomed. If the movement of married women into the labor force proceeded with what some social critics regarded as unseemly speed, it did so because many members of the generation that had been raised in the affluent society (those who were children of the baby boom, not of the Depression, and who came to maturity and began forming their households in the 1960s and 1970s) saw little reason to worry about the various social ills that might result from cold cereal for breakfast, from an occasional meal in a restaurant, from slightly dirty bathroom sinks and unironed sheets. Modern household technology facilitated married women's workforce participation not by freeing women from household labor but by making it possible for women to maintain decent standards in their homes without assistants and without a full-time commitment to house-work.

CONCLUSION

The work that women do when they are being paid to do it is easy to recognize, because there are so many standard indicators that allow us to account for it—personnel records, time clocks, pay sheets, and the like. On the other hand, the productive labor that is still being done in American homes is difficult to recognize, because the reigning theory of family history tells us that it should not be there, because the reigning methodology of the social sciences cannot be applied to it, because ordinary language has a penchant for masking it, and because advertisers have had a vested interest in convincing us that it has evaporated. Economists and sociologists do not consider housework to be "productive work," at least in part because they cannot measure it. They can easily quantify what people are consuming (how many cans of peas? how many dollars' worth of stockings?), but they cannot place a dollar value (to choose a particularly simple example) on a nutritious meal— and they cannot begin to estimate how many such meals are prepared in households throughout the year (in part, because the workers who prepare them are not paid nor are their hours timed). People who write advertising copy for microwave ovens, toilet bowl cleaners, and paper toweling seem to believe that they will lose their jobs if they confess that it still takes time to prepare food for the oven, scrub the brown stains out of the toilet, and wipe down counters after dinner has been consumed. Virtually every lecture on the history of the family, and every textbook on the sociology of the family, and every new inquiry into the state of the family begins with the sentiment that "households do not produce anything valuable any more." And, in our everyday conversations, we cannot even refer to housewives as "laboring" or as "working" or even as being "employed," without confusing our listeners, even though we all know that housework is work.

The technological systems that presently dominate our households were built on the assumption that a full-time housewife would be operating them, since very few people in the last one hundred years (when the foundations for these systems were being laid) wanted adult women to leave their homes in order to work in the labor market, or believed that adult women themselves would ever want to go out to work. In the earliest stages of industrialization, in the early decades of the nineteenth century, as some of men's work in the home was eliminated (fuel gathering, leather working, grain processing), some men were thereby freed to work (at least part of the year) in factories and offices. Some of the women's housework was eliminated at that time also (principally spinning and weaving), but no one then expected or desired women to leave their homes to work for wages elsewhere (unless the women were single or exceedingly poor) because so much of what had always been considered women's work still remained to be done at home: cooking, sewing, laundering, cleaning, child care. In the next stages of industrialization, even more of men's household work was eliminated, as was much of children's work; but, again, no one expected or desired women to leave their homes in order to go out to work because, whether rich or poor, a family's sustenance and status still depended on the presence of a full-time homemaker. In this stage of industrialization (roughly from 1880 to 1920), the foundations for the modern household technologies were laid: municipalities began to supply households with clean water and ample sewers; gas and electric companies figured out how to bring in modern fuels; merchandisers and retailers developed new techniques for selling durable goods to households. Almost no one who participated in this process—whether rich or poor, whether female or male, whether producer or consumer—seems to have doubted that the individual household would be the ultimate consumption unit, and that most of the work of that household would be done by housewives who would continue to work, as they had in the past, without pay and without timeclocks. If the utility companies had had any reason to believe that households would stop functioning after five or six o'clock—as offices, stores, and many factories do—they would have had precious little motivation for trying to supply them with electricity, water, and gas. Similarly, if householders had believed that they would have to pay every adult woman for every hour that she labored in their homes, they would have had precious little economic incentive for preferring washing machines to commercial laundry services and automobiles to deliverymen. Whether for good or ill, women were the only workers whose "place" was still at home in the years when homes were becoming mechanized, and the vast majority of these women were housewives who were not paid hourly, weekly, or even annual wages. When, in the decades after the Second World War, our economy finally became capable of realizing the potential benefits of these technological systems, the individual household, the individual ownership of tools, and the allocation of housework to women had, almost literally, been cast in the stainless steel, the copper, and the aluminum out of which those systems were composed.

The implications of this arrangement and the ironies implicit in it became particularly clear to those millions upon millions of families who moved out of urban areas and into suburban ones in the postwar decades. The move to the suburbs carried with it the assumption that someone (surely mother) would be at home to do the requisite work that made it possible for someone else (surely father) to leave early in the morning and return late at night, without worrying either about the welfare of his family or the maintenance of his domicile. Having made the move and purchased the house and invested in the cars and the appliances without which the suburban way of life simply was not possible, people discovered that the technological systems in which they had invested (not only so much money, but also so much emotion) simply would not function unless someone stayed home to operate them.

When this "someone" had, however, decided that, for whatever reason, staying at home was no longer her cup of tea, neither the house nor the cars nor the appliances nor the way of life that they all implied could simply be thrown into the dustbin, nor did anyone wish to throw them there. All of these were long-term investments (consumer *durables*); and the technological systems of which they were a part (houses, roads, telephone lines, gas mains) were built to last for more than one lifetime. The transition to the two-income family (or to the female-headed household) did not occur without taking a toll—a toll measured in the hours that employed housewives had to work in order to perform adequately first as employees and then as housewives. A thirty-five-hour week (housework) added to a forty-hour week (paid employment) adds up to a working week that even sweatshops cannot match. With all her appliances and amenities, the status of being a "working mother" in the United States today is, as three eminent experts have suggested, virtually a guarantee of being overworked and perpetually exhausted.

The technological and social systems for doing housework had been constructed with the expectation that the people engaged in them would be full-time housewives. When the full-time housewives began to disappear, those systems could not adjust quickly. Not even the most efficient working wife in the world can prepare, serve, and clean up from a meal in four minutes flat; and even the best organized working mother still cannot feed breakfast to a toddler in thirty seconds. Homes cannot automatically be moved close to a job or even close to public transportation, so someone still has to be available to drive the man of the family to the train or a child to the soccer field or to a party; and day-care centers cannot quickly be built where they have not existed before, so someone still has to leave a career behind for a while when babies are born—or find a helpful grandmother.

Indeed, given the sacred feelings that most Americans seem to attach to meals, infants, private homes, and clean laundry—and given the vast investment individuals, corporations, and municipalities have made in the technological systems that already exist—our household technologies may never evolve so as to make life easier for the working wife and mother. In the generations to come, housework is not likely to disappear. Barring a catastrophic economic or nuclear disaster, the vast majority of today's children will form families when they grow up, will buy houses, and will outfit those houses with tools for doing housework. Home computers may be added to the repertoire, but there will still be at least functional equivalents of cooking stoves and refrigerators, telephones and automobiles, washing machines and dishwashers. However much trouble these technologies may be, however much they may cost to obtain and then to maintain, and however much they may induce us to engage in amounts or forms of work that are often irritating and sometimes infuriating, the standard of living and the way of living that they make possible is one to which many Americans aspired in the past and that many are unlikely to forsake in the future. The washing machine may not save as much time as its advertisers might like us to believe, and electricity may not bring as many good things to living as the manufacturers of generating equipment would like us to think, but the daily lives that are shaped by washing machines and electricity are so much more comfortable and healthy than the ones that were shaped by washtubs and coal (or, before that, dirty clothes and open hearths) that we will probably not give them up.

Still, while enjoying the benefits that these technological systems provide, we need not succumb entirely to the work processes that they seem to have ordained for us. If we regard these processes as unsatisfactory, we can begin to extricate ourselves from them not by destroying the technological systems with which they are associated but by revising the unwritten rules that govern the systems. Some of these rules—to change our sheets once a week and keep our sinks spotless and greaseless, to wipe the table after every meal, to flush the toilet, brush our teeth, change our clothes and wash our hair, to give music lessons to our children and keep our dirty linen literally and figuratively to ourselves—generate more housework than may really be necessary. These rules were passed down to us by members of an earlier generation (our parents) and sprang from fear of the deprivations that poverty engenders and from a desire either to rise above those deprivations or to stave them off. Now that profound poverty has ceased to be an imminent threat for most of us, the time has surely come to re-evaluate the amount of time that we spend maintaining the symbols of our status.

Others of these rules—that, for example, men who dry dishes or change diapers are insufficiently masculine, that only women can properly nurture infants, that young girls should help their mothers in the kitchen and young boys assist their fathers in the garage, that husbands can undertake long commutes but wives cannot—ensure that the work processes of housework will be confined to members of only one sex, not only in this generation but in generations to come. These latter rules, connected as they are to aspects of our sexuality and our self-conception, are not easy to revise. Even those brave members of the postwar generations who learned to sever the bond between "working mother" and "social disaster" could not erase more than one social stereotype at a time; and when they chose spouses and formed households, they adopted virtually the same sexual division of household labor with washing machines and microwave ovens as had their ancestors with washtubs and open hearths: the men responsible for fuel and for lawns (those symbolic remnants of fields of waving grain) and the women responsible for cooking, cleaning, laundering, and child care. If centuries upon centuries of social conditioning have led us to prefer the private household and the individual ownership of tools, then centuries upon centuries of social conditioning also prepared these young women to be housewives and these young men to believe that the work of cooking, cleaning, and caring for infants would threaten their masculinity. Indeed, when the children of the baby boom were still children, when they were forming their sense of "what it means to be a woman" or "what it means to be a man," all the adults upon whom these adolescents might have been modeling themselves—their parents, the people down the block, celebrities, creators of plots for movies, authors of magazine articles and textbooks—were still engaged in the backward search for femininity and still suggesting (in the strongest affective terms) that dishwashers and diapers were objects to be manipulated by females, and that wrenches and lawnmowers were objects to be manipulated by males, and that the manipulation of inappropriate objects was, to put it anthropologically, sexually polluting.

The rules that stem from a fear of poverty, and the rules that stem from fear of sexual pollution, were the product of specific historical periods, with social and technological constraints of their own. The widespread diffusion of modern household technology and the widespread entrance of married women into the labor force have markedly loosened those constraints; and thus the time has come to begin changing the rules. We can best solve the problems that beset many working wives and their families not by returning to the way things used to be (since that is probably impossible and, in view of the way things really used to be, hardly attractive), not by destroying the technological systems that have provided many benefits (and that much of the rest of the world is trying, for fairly good reasons, to emulate), and not by calling for the death of the family as a social institution (a call that the vast majority of people are unlikely to heed)—but by helping the next generation (and ourselves) to neutralize both the sexual connotation of washing machines and vacuum cleaners and the senseless tyranny of spotless shirts and immaculate floors.

CHAPTER 14

Amusing Ourselves to Death: Public Discourse in the Age of Show Business

Neil Postman

Is there any technological development in the twentieth century that has had a more visible impact on American society than television? The ubiquitous television set has profoundly changed news reporting, political campaigning, habits of consumption, and the ways in which we raise children—to name only a few of the often mentioned, and also often lamented—aspects of its influence.

Neil Postman is one of the many critics who has raised concerns about the television's negative impact on our society. But he stands apart from the other critics for his novel approach: he does not criticize television's output of junk. Entertainment is fine, he says. But unlike other entertaining media, television's influence has invaded the more serious, higher realm of our "public discourse," our society's discussion of public affairs; consequently television has reduced *everything* to entertainment.

In the selection presented here, "The Age of Show Business," Postman provides examples of what he regards to be alarming evidence of television's pernicious influence throughout society. Television is not, he says, an extension of the preceding literate tradition, that is, the tradition based on the printed word that television has supplanted. To Postman, television is an entirely different animal, and different on two separate levels.

First, the very technology of television, like all technology, contains a bias that will predispose it to be used in certain ways and not in others. Second, the television technology creates a social and intellectual environment in the particular place it is used—this Postman calls television as a *medium*—and this too will pull a technology's applications in a certain direction. Postman attaches more significance to this second level of social application than to the first, observing that if television is understood as nothing but the bare technology, then it should have had a similar impact in socialist and capitalist countries alike, which it most certainly has not.

Here in the United States, television's potential as a technology of visual images has fully flowered. But at the same time, however, television has insisted that all of its programming be an entertainment. Postman mocks television's attempts to provide more serious content, such as contemporary television news programs and a televised panel "discussion" following the broadcast of "The Day After," the chilling movie portraying nuclear holocaust.

Worse, Postman maintains, television sets an entertainment standard that people have become accustomed to use even after they have turned the television set off. Public affairs—politics, religion, business, education, law, and other serious matters—formerly were conducted with the print medium serving as the dominating medium, what Postman calls the "media-metaphor." But now television is the dominant media-metaphor and no area of American life seems immune from television's commanding imperative that we always be entertained. Postman provides illustrations: the disc jockey priest and joking archbishop; televised trials of rape defendants; Meryl Streep upstaging Mother Teresa at Yale's commencement ceremonies; 1984 presidential candidate debates that were not debates in any meaningful sense; and other examples.

Postman argues that the line separating show business from what is not show business has become increasingly indiscernible. Perhaps still one other example of the trend should be added here: the commercial success of the national newspaper *U.S.A. Today*. Here we have a remnant of the bygone era of print—the daily newspaper—surviving by making itself as much like television as it can: short, snappy, upbeat stories (the paper has been dubbed "McPaper" for its bland, reliable servings, like McDonald's); splashy colors and graphics that please the eye more than the mind; strong sports coverage; and weak international coverage.

What should be done? The critics can have their say, but the general public seems to have cast its vote decisively in favor of having its news served light and easy. "Eyewitness News" is preferred to the "MacNeil-Lehrer News Hour,"; *U.S.A. Today* to the *New York Times*. The present does not seem to offer indications of likely changes in this pattern in the foreseeable future. Postman and like-minded critics must contend with a democratic tradition that enshrines the people's choice as the best.

THE AGE OF SHOW BUSINESS

A dedicated graduate student I know returned to his small apartment the night before a major examination only to discover that his solitary lamp was broken beyond repair. After a whiff of panic, he was able to restore both his equanimity and his chances for a satisfactory grade by turning on the television set, turning off the sound, and with his back to the set, using its light to read important passages on which he was to be tested. This is one use of television—as a source of illuminating the printed page.

But the television screen is more than a light source. It is also a smooth, nearly flat surface on which the printed word may be displayed. We have all stayed at hotels in which the TV set has had a special channel for describing the day's events in letters rolled endlessly across the screen. This is another use of television—as an electronic bulletin board.

Many television sets are also large and sturdy enough to bear the weight of a small library. The top of an old-fashioned RCA console can handle as many as thirty books, and I know one woman who has securely placed her entire collection of Dickens, Flaubert, and Turgenev on the top of a 21-inch Westinghouse. Here is still another use of television—as bookcase.

I bring forward these quixotic uses of television to ridicule the hope harbored by some that television can be used to support the literate tradition. Such a hope represents exactly what Marshall McLuhan used to call "rear-view mirror" thinking: the assumption that a new medium is merely an extension or amplification of an older one; that an automobile, for example, is only a fast horse, or an electric light a powerful candle. To make such a mistake in the matter at hand is to misconstrue entirely how television redefines the meaning of public discourse. Television does not extend or amplify literate culture. It attacks it. If television is a continuation of anything, it is of a tradition begun by the telegraph and photograph in the mid-nineteenth century, not by the printing press in the fifteenth.

What is television? What kinds of conversations does it permit? What are the intellectual tendencies it encourages? What sort of culture does it produce?

These are the questions to be addressed in the rest of this book, and to approach them with a minimum of confusion, I must begin by making a distinction between a technology and a medium. We might say that a technology is to a medium as the brain is to the mind. Like the brain, a technology is a physical apparatus. Like the mind, a medium

SOURCE: From *Amusing Ourselves to Death* by Neil Postman. Copyright © 1985 by Neil Postman. Reprinted by permission of Viking Penguin, Inc.

is a use to which a physical apparatus is put. A technology becomes a medium as it employs a particular symbolic code, as it finds its place in a particular social setting, as it insinuates itself into economic and political contexts. A technology, in other words, is merely a machine. A medium is the social and intellectual environment a machine creates.

Of course, like the brain itself, every technology has an inherent bias. It has within its physical form a predisposition toward being used in certain ways and not others. Only those who know nothing of the history of technology believe that a technology is entirely neutral. There is an old joke that mocks that naive belief. Thomas Edison, it goes, would have revealed his discovery of the electric light much sooner than he did except for the fact that every time he turned it on, he held it to his mouth and said, "Hello? Hello?"

Not very likely. Each technology has an agenda of its own. It is, as I have suggested, a metaphor waiting to unfold. The printing press, for example, had a clear bias toward being used as a linguistic medium. It is *conceivable* to use it exclusively for the reproduction of pictures. And, one imagines, the Roman Catholic Church would not have objected to its being so used in the sixteenth century. Had that been the case, the Protestant Reformation might not have occurred, for as Luther contended, with the word of God on every family's kitchen table, Christians do not require the Papacy to interpret it for them. But in fact there never was much chance that the press would be used solely, or even very much, for the duplication of icons. From its beginning in the fifteenth century, the press was perceived as an extraordinary opportunity for the display and mass distribution of written language. Everything about its technical possibilities led in that direction. One might even say it was invented for that purpose.

The technology of television has a bias, as well. It is conceivable to use television as a lamp, a surface for texts, a bookcase, even as radio. But it has not been so used and will not be so used, at least in America. Thus, in answering the question, What is television?, we must understand as a first point that we are not talking about television as a technology but television as a medium. There are many places in the world where television, though the same technology as it is in America, is an entirely different medium from that which we know. I refer to places where the majority of people do not have television sets, and those who do have only one; where only one station is available; where television does not operate around the clock; where most programs have as their purpose the direct furtherance of government ideology and policy; where commercials are unknown, and "talking heads" are the principal image; where television is mostly used as if it were radio. For these reasons and more television will not have the same meaning or power as it does in America, which is to say, it is possible for a technology to be so used that its potentialities are prevented from developing and its social consequences kept to a minimum.

But in America, this has not been the case. Television has found in liberal democracy and a relatively free market economy a nurturing climate in which its full potentialities as a technology of images could be exploited. One result of this has been that American television programs are in demand all over the world. The total estimate of U.S. television program exports is approximately 100,000 to 200,000 hours, equally divided among Latin America, Asia and Europe. Over the years, programs like "Gunsmoke," "Bonanza," "Mission: Impossible," "Star Trek," "Kojak," and more recently, "Dallas" and "Dynasty" have been as popular in England, Japan, Israel and Norway as in Omaha, Nebraska. I have heard (but not verified) that some years ago the Lapps postponed for several days their annual and, one supposes, essential migratory journey so that they could find out who shot J.R. All of this has occurred simultaneously with the decline of America's moral and political prestige, worldwide. American television programs are in demand not because America is loved but because American television is loved.

We need not be detained too long in figuring out why. In watching American television, one is reminded of George Bernard Shaw's remark on his first seeing the glittering neon signs of Broadway and 42nd Street at night. It must be beautiful, he said, if you cannot read. American television is, indeed, a beautiful spectacle, a visual delight, pouring forth thousands of images on any given day. The average length of a shot on network television is only 3.5 seconds, so that the eye never rests, always has something new to see. Moreover, television offers viewers a variety of subject matter, requires minimal skills to comprehend it, and is largely aimed at emotional gratification. Even commercials, which some regard as an annoyance, are exquisitely crafted, always pleasing to the eye and accompanied by exciting music. There is no question but that the best photography in the world is presently seen on television commercials. American television, in other words, is devoted entirely to supplying its audience with entertainment.

Of course, to say that television is entertaining is merely banal. Such a fact is hardly threatening to a culture, not even worth writing a book about. It may even be a reason for rejoicing. Life, as we like to say, is not a highway strewn with flowers. The sight of a few blossoms here and there may make our journey a trifle more endurable. The Lapps undoubtedly thought so. We may surmise that the ninety million Americans who watch television every night also think so. But what I am claiming here is not that television is entertaining but that it has made entertainment itself the natural format for the representation of all experience. Our television set keeps us in constant communion with the world, but it does so with a face whose smiling countenance is unalterable. The problem is not that television presents us with entertaining subject matter but that all subject matter is presented as entertaining, which is another issue altogether.

To say it still another way: Entertainment is the supraideology of all discourse on television. No matter what is depicted or from what point of view, the overarching presumption is that it is there for our amusement and pleasure. That is why even on news shows which provide us daily with fragments of tragedy and barbarism, we are urged by the newscasters to "join them tomorrow." What for? One would think that several minutes of murder and mayhem would suffice as material for a month of sleepless nights. We accept the newscasters' invitation because we know that the "news" is not to be taken seriously, that it is all in fun, so to say. Everything about a news show tells us this—the good looks and amiability of the cast, their pleasant banter, the exciting music that opens and closes the show, the vivid film footage, the attractive commercials—all these and more suggest that what we have just seen is no cause for weeping. A news show, to put it plainly, is a format for entertainment, not for education, reflection or catharsis. And we must not judge too harshly those who have framed it in this way. They are not assembling the news to be read, or broadcasting it to be heard. They are televising the news to be seen. They must follow where their medium leads. There is no conspiracy here, no lack of intelligence, only a straightforward recognition that "good television" has little to do with what is "good" about exposition or other forms of verbal communication but everything to do with what the pictorial images look like.

I should like to illustrate this point by offering the case of the eighty-minute discussion provided by the ABC network on November 20, 1983, following its controversial movie *The Day After*. Though the memory of this telecast has receded for most, I choose this case because, clearly, here was television taking its most "serious" and "responsible" stance. Everything that made up this broadcast recommended it as a critical test of television's capacity to depart from an entertainment mode and rise to the level of public instruction. In the first place, the subject was the possibility of a nuclear holocaust. Second, the film itself had been attacked by several influential bodies politic, including the Reverend Jerry Falwell's Moral Majority. Thus, it was important that the network display television's value and serious intentions as a medium of information and coherent discourse. Third, on the program itself no musical theme was used as background—a significant point since almost all television programs are embedded in music, which helps to tell the audience what emotions are to be called forth. This is a standard theatrical device, and its absence on television is always ominous. Fourth, there were no commercials during the discussion, thus elevating the tone of the event to the state of reverence usually reserved for the funerals of assassinated Presidents. And finally, the participants included Henry Kissinger, Robert McNamara, and Elie Wiesel, each of whom is a symbol of sorts of serious discourse. Although Kissinger, somewhat later, made an appearance on the hit show "Dynasty," he was then and still is a paradigm of intellectual sobriety; and Wiesel, practically a walking metaphor of social conscience. Indeed, the other members of the cast—Carl Sagan, William Buckley and General Brent Scowcroft—are, each in his way, men of intellectual bearing who are not expected to participate in trivial public matters.

The program began with Ted Koppel, master of ceremonies, so to speak, indicating that what followed was not intended to be a debate but a *discussion*. And so those who are interested in philosophies of discourse had an excellent opportunity to observe what serious television means by the word "discussion." Here is what it means: Each of six men was given approximately five minutes to say something about the subject. There was, however, no agreement on exactly what the subject was, and no one felt obliged to respond to anything anyone else said. In fact, it would have been difficult to do so, since the participants were called upon seriatim, as if they were finalists in a beauty contest, each being given his share of minutes in front of the camera. Thus, if Mr. Wiesel, who was called upon last, had a response to Mr. Buckley, who was called upon first, there would have been four commentaries in between, occupying about twenty minutes, so that the audience (if not Mr. Wiesel himself) would have had difficulty remembering the argument which prompted his response. In fact, the participants—most of whom were no strangers to television—largely avoided addressing each other's points. They used their initial minutes and then their subsequent ones to intimate their position or give an impression. Dr. Kissinger, for example, seemed intent on making viewers feel sorry that he was no longer their Secretary of State by reminding everyone of books he had once written, proposals he had once made, and negotiations he had once conducted. Mr. McNamara informed the audience that he had eaten lunch in Germany that very afternoon, and went on to say that he had at least fifteen proposals to reduce nuclear arms. One would have thought that the discussion would turn on this issue, but the others seemed about as interested in it as they were in what he had for lunch in Germany. (Later, he took the initiative to mention three of his proposals but they were not discussed.) Elie Wiesel, in a series of quasi-parables and paradoxes, stressed the tragic nature of the human condition, but because he did not have the time to provide a context for his remarks, he seemed quixotic and confused, conveying an impression of an itinerant rabbi who has wandered into a coven of Gentiles.

In other words, this was no discussion as we normally use the word. Even when the "discussion" period began, there were no arguments or counterarguments, no scrutiny of assumptions, no explanations, no elaborations, no definitions. Carl Sagan made, in my opinion, the most coherent statement—a four-minute rationale for a nuclear freeze—but it contained at least two questionable assumptions and was not carefully examined. Apparently, no one wanted to take time from his own few minutes to call attention to someone else's. Mr. Koppel, for his part, felt obliged to keep the "show" moving, and though he occasionally pursued what he discerned as a line of thought, he was more concerned to give each man his fair allotment of time.

But it is not time constraints alone that produce such fragmented and discontinuous language. When a television show is in process, it is very nearly impermissible to say, "Let me think about that" or "I don't know" or "What do you mean when you say. . . ?" or "From what sources does your information come?" This type of discourse not only slows down the tempo of the show but creates the impression of uncertainty or lack of finish. It tends to reveal people in the *act of thinking*, which is as disconcerting and boring on television as it is on a Las Vegas stage. Thinking does not play well on television, a fact that television directors discovered long ago. There is not much to *see* in it. It is, in a phrase, not a performing art. But television demands a performing art, and so what the ABC network gave us was a picture of men of sophisticated verbal skills and political understanding being brought to heel by a medium that requires them to fashion performances rather than ideas. Which acccounts for why the eighty minutes were very entertaining, in the way of a Samuel Beckett play: The intimations of gravity hung heavy, the meaning passeth all understanding. The performances, of course, were highly professional. Sagan abjured the turtle-neck sweater in which he starred when he did "Cosmos." He even had his hair cut for the event. His part was that of the logical scientist speaking in behalf of the planet. It is to be doubted that Paul Newman could have done better in the role, although Leonard Nimoy might have. Scowcroft was suitably military in his bearing—terse and distant, the unbreakable defender of national security. Kissinger, as always, was superb in the part of the knowing world statesman, weary of the sheer responsibility of keeping disaster at bay. Koppel played to perfection the part of a moderator, pretending, as it were, that he was sorting out ideas while, in fact, he was merely directing the performances. At the end, one could only applaud those performances, which is what a good television program always aims to achieve; that is to say, applause, not reflection.

I do not say categorically that it is impossible to use television as a carrier of coherent language or thought in process. William Buckley's own program, "Firing Line," occasionally shows people in the act of thinking but who also happen to have television cameras pointed at them. There are other programs, such as "Meet the Press" or "The Open Mind," which clearly strive to maintain a sense of intellectual decorum and typographic tradition, but they are scheduled so that they do not compete with programs of great visual interest, since otherwise, they will not be watched. After all, it is not unheard of that a format will occasionally go against the bias of its medium. For example, the most popular radio program of the early 1940's featured a ventriloquist, and in those days, I heard more than once the feet of a tap dancer on the "Major Bowes' Amateur Hour." (Indeed, if I am not mistaken, he even once featured a pantomimist.) But ventriloquism, dancing and mime do not play well on radio, just as sustained, complex talk does not play well on television. It can be made to play tolerably well if only one camera is used and the visual image is kept constant—as when the President gives a speech. But this is not television at its best, and it is not television that most people will choose to watch. The single most important fact about television is that people *watch* it, which is why it is called "*television*." And what they watch, and like to watch, are moving pictures—millions of them, of short duration and dynamic variety. It is in the nature of the medium that it must suppress the content of ideas in order to accommodate the requirements of visual interest; that is to say, to accommodate the values of show business.

Film, records and radio (now that it is an adjunct of the music industry) are, of course, equally devoted to entertaining the culture, and their effects in altering the style of American discourse are not insignificant. But television is different because it encompasses all forms of discourse. No one goes to a movie to find out about government policy or the latest scientific advances. No one buys a record to find out the baseball scores or the weather or the latest murder. No one turns on radio anymore for soap operas or a presidential address (if a television set is at hand). But everyone goes to television for all these things and more, which is why television resonates so powerfully throughout the culture. Television is our culture's principal mode of knowing about itself. Therefore—and this is the critical point—how television stages the world becomes the model for how the world is properly to be staged. It is not merely that on the television screen entertainment is the metaphor for all discourse. It is that off the screen the same metaphor prevails. As typography once dictated the style of conducting politics, religion, business, education, law and other important social matters, television now takes command. In courtrooms, classrooms, operating rooms, board rooms, churches and even airplanes, Americans no longer talk to each other, they entertain each other. They do not exchange ideas; they exchange images. They do not argue with propositions; they argue with good looks, celebrities and commercials. For the message of television as metaphor is not only that all the world is a stage but that the stage is located in Las Vegas, Nevada.

In Chicago, for example, the Reverend Greg Sakowicz, a Roman Catholic priest, mixes his religious teaching with rock 'n' roll music. According to the Associated Press, the Reverend Sakowicz is both an associate pastor at the Church of the Holy Spirit in Schaumberg (a suburb of Chicago) and a disc jockey at WKQX. On his show, "The Journey Inward," Father Sakowicz chats in soft tones about such topics as family relationships or commitments, and interposes his sermons with "the sound of *Billboard's* Top 10." He says that his preaching is not done "in a churchy way," and adds, "You don't have to be boring in order to be holy."

Meanwhile in New York City at St. Patrick's Cathedral, Father John J. O'Connor put on a New York Yankee baseball cap as he mugged his way through his installation as Archbishop of the New York Archdiocese. He got off some excellent gags, at least one of which was specifically directed at Mayor Edward Koch, who was a member of his audience; that is to say, he was a congregant. At his next public performance, the new archbishop donned a New York Mets baseball cap. These events were, of course, televised, and were vastly entertaining, largely because Archbishop (now Cardinal) O'Connor has gone Father Sakowicz one better: Whereas the latter believes that you don't have to be boring to be holy, the former apparently believes you don't have to be holy at all.

In Phoenix, Arizona, Dr. Edward Dietrich performed triple bypass surgery on Bernard Schuler. The operation was successful, which was nice for Mr. Schuler. It was also on television, which was nice for America. The operation was carried by at least fifty television stations in the United States, and also by the British Broadcasting Corporation. A two-man panel of narrators (a play-by-play and color man, so to speak) kept viewers informed about what they were seeing. It was not clear as to why this event was televised, but it resulted in transforming both Dr. Dietrich and Mr. Schuler's chest into celebrities. Perhaps because he has seen too many doctor shows on television, Mr. Schuler was uncommonly confident about the outcome of his surgery. "There is no way in hell they are going to lose me on live TV," he said.

As reported with great enthusiasm by both WCBS-TV and WNBC-TV in 1984, the Philadelphia public schools have embarked on an experiment in which children will have their curriculum sung to them. Wearing Walkman equipment, students were shown listening to rock music whose lyrics were about the eight parts of speech. Mr. Jocko Henderson, who thought of this idea, is planning to delight students further by subjecting mathematics and history, as well as English, to the rigors of a rock music format. In fact, this is not Mr. Henderson's idea at all. It was pioneered by the Children's Television Workshop, whose television show "Sesame Street" is an expensive illustration of the idea that education is indistinguishable from entertainment. Nonetheless, Mr. Henderson has a point in his favor. Whereas "Sesame Street" merely attempts to make learning to read a form of light entertainment, the Philadelphia experiment aims to make the classroom itself into a rock concert.

In New Bedford, Massachusetts, a rape trial was televised, to the delight of audiences who could barely tell the difference between the trial and their favorite mid-day soap opera. In Florida, trials of varying degrees of seriousness, including murder, are regularly televised and are considered to be more entertaining than most fictional courtroom dramas. All of this is done in the interests of "public education." For the same high purpose, plans are afoot, it is rumored, to televise confessionals. To be called "Secrets of the Confessional Box," the program will, of course, carry the warning that some of its material may be offensive to children and therefore parental guidance is suggested.

On a United Airlines flight from Chicago to Vancouver, a stewardess announces that its passengers will play a game. The passenger with the most credit cards will win a bottle of champagne. A man from Boston with twelve credit cards wins. A second game requires the passengers to guess the collective age of the cabin crew. A man from Chicago guesses 128, and wins another bottle of wine. During the second game, the air turns choppy and the Fasten Seat Belt sign goes on. Very few people notice, least of all the cabin crew, who keep up a steady flow of gags on the intercom. When the plane reaches its destination, everyone seems to agree that it's fun to fly from Chicago to Vancouver.

On February 7, 1985, *The New York Times* reported that Professor Charles Pine of Rutgers University (Newark campus) was named Professor of the Year by the Council for the Support and Advancement of Education. In explaining why he has such a great impact on his students, Professor Pine said: "I have some gimmicks I use all the time. If you reach the end of the blackboard, I keep writing on the wall. It always gets a laugh. The way I show what a glass molecule does is to run over to one wall and bounce off it, and run over to the other wall." His students are, perhaps, too young to recall that James Cagney used this "molecule move" to great effect in *Yankee Doodle Dandy*. If I am not mistaken, Donald O'Connor duplicated it in *Singin' in the Rain*. So far as I know, it has been used only once before in a classroom: Hegel tried it several times in demonstrating how the dialectical method works.

The Pennsylvania Amish try to live in isolation from mainstream American culture. Among other things, their religion opposes the veneration of graven images, which means that the Amish are forbidden to see movies or to be photographed. But apparently their religion has not got around to disallowing seeing movies *when* they are being photo-graphed. In the summer of 1984, for example, a Paramount Pictures crew descended upon Lancaster County to film the movie *Witness,* which is about a detective, played by Harrison Ford, who falls in love with an Amish woman. Although the Amish were warned by their church not to interfere with the film makers, it turned out that some Amish welders ran to see the action as soon as their work was done. Other devouts lay in the grass some distance away, and looked down on the set with binoculars. "We read about the movie in the paper," said an Amish woman. "The kids even cut out Harrison Ford's picture." She added: "But it doesn't really matter that much to them. Somebody told us he was in *Star Wars* but that doesn't mean anything to us." The last time a similar conclusion was drawn was when the executive director of the American Association of Blacksmiths remarked that he had read about the automobile but that he was convinced it would have no consequences for the future of his organization.

In the Winter, 1984, issue of the *Official Video Journal* there appears a full-page advertisement for "The Genesis Project." The project aims to convert the Bible into a series of movies. The end-product, to be called "The New Media Bible," will consist of 225 hours of film and will cost a quarter of a billion dollars. Producer John Heyman, whose credits include *Saturday Night Fever* and *Grease,* is one of the film makers most committed to the project. "Simply stated," he is quoted as saying, "I got hooked on the Bible." The famous Israeli actor Topol, best known for his role of Tevye in *Fiddler on the Roof,* will play the role of Abraham. The advertisement does not say who will star as God but, given the producer's background, there is some concern that it might be John Travolta.

At the commencement exercises at Yale University in 1983, several honorary degrees were awarded, including one to Mother Teresa. As she and other humanitarians and scholars, each in turn, received their awards, the audience applauded appropriately but with a slight hint of reserve and impatience, for it wished to give its heart to the final recipient who waited shyly in the wings. As the details of her achievements were being recounted, many people left their seats and surged toward the stage to be closer to the great woman. And when the name Meryl Streep was announced, the audience unleashed a sonic boom of affection to wake the New Haven dead. One man who was present when Bob Hope received his honorary doctorate at another institution said that Dr. Streep's applause surpassed Dr. Hope's. Knowing how to please a crowd as well as anyone, the intellectual leaders at Yale invited Dick Cavett, the talk-show host, to deliver the commencement address the following year. It is rumored that this year, Don Rickles will receive a Doctorate of Humane Letters and Lola Falana will give the commencement address.

Prior to the 1984 presidential elections, the two candidates confronted each other on television in what were called "debates." These events were not in the least like the Lincoln-Douglas debates or anything else that goes by the name. Each candidate was given five minutes to address such questions as, What is (or would be) your policy in Central America? His opposite number was then given one minute for a rebuttal. In such circumstances, complexity, documentation and logic can play no role, and, indeed, on several occasions syntax itself was abandoned entirely. It is no matter. The men were less concerned with giving arguments than with "giving off" impressions, which is what television does best. Post-debate commentary largely avoided any evaluation of the candidates' ideas, since there were none to evaluate. Instead, the debates were conceived as boxing matches, the relevant question being, Who KO'd whom? The answer was determined by the "style" of the men—how they looked, fixed their gaze, smiled, and delivered one-liners. In the second debate, President Reagan got off a swell one-liner when asked a question about his age. The following day, several newspapers indicated that Ron had KO'd Fritz with his joke. Thus, the leader of the free world is chosen by the people in the Age of Television.

What all of this means is that our culture has moved toward a new way of conducting its business, especially its important business. The nature of its discourse is changing as the demarcation line between what is show business and what is not becomes harder to see with each passing day. Our priests and presidents, our surgeons and lawyers, our educators and newscasters need worry less about satisfying the demands of their discipline than the demands of good showmanship. Had Irving Berlin changed one word in the title of his celebrated song, he would have been as prophetic, albeit more terse, as Aldous Huxley. He need only have written, There's No Business But Show Business.

CHAPTER 15

The Second Self: Computers and the Human Spirit

Sherry Turkle

The personal computer is a product of technology that is too young to claim the title of having had the most impact on twentieth-century American society, but some believe that it will soon prove to be more influential in important ways than even television. Sherry Turkle grew interested in how the personal computer affects how we think of ourselves, one another, and our relationship with the world. Her book *The Second Self* records the discoveries made during Turkle's interviews with 400 children and adults who were working with computers in a period when the machines were still novelties (1977–1983).

Turkle believes that any technology is basically blank, that it does *not* possess inherent predispositions to be used certain ways instead of others. In her view, the computer is inert and serves merely as a mirror that reveals what humans bring to it. In her book, she shows that the course of computer technology's impact on society is not on one single "society," but on many different societies, such as girls and boys, children and adults, computer-fanatic hackers and computer-phobic novices.

In this selection, taken from her chapter on "Child Programmers," she tells us about her research at a private school referred to as Austen (a pseudonym) where children were given liberal access to computers.

The very idea that young children would be encouraged to spend time with these machines, in the case of some even before they had learned how to write with a pencil, could be unsettling. Turkle anticipates objections that could be raised—that children should not be kept indoors interacting with a machine instead of romping outside, and that children should not lose their childhood innocence prematurely. But she claims that the machines have no observably universal effect upon the children who use them. She regards the interesting question not to be what does the computer do to the children, but rather what do different kinds of children make of the machine.

While observing how second, third, and fourth graders went about learning how to program using the graphics-oriented language Logo, Turkle saw different approaches spontaneously taken by the students, different "styles of mastery," as they explored the machine's capabilities largely on their own. "Hard" masters were expert planners and were intent on imposing their will upon the machine. "Soft" masters were less intent on carrying out a set design and were more reactive, changing their plans as they went along. Turkle maintains that the differences in programming styles that she saw in the individual children merely reflected the deeper differences in personality: the "hards" were more comfortable operating on inanimate objects in general, and the "softs" were more socially oriented and more compromising by nature.

Girls tend to be softs, and boys hards. Why this should be so is a question for which a number of theories have been suggested. Turkle addresses two explanations. First, the socialization explanation points to the differences in how boys and girls are raised in our culture. Boys are encouraged to command, direct, build, and play with things, while girls are taught the characteristics of soft mastery—negotiation, compromise, give-and-take. Second, the psychoanalytic explanation, to which Turkle devotes most of her attention, points to the supposition that boys pass through a traumatic Oedipal stage early in their psychological development that makes them emotionally cooler. Here Turkle introduces the term *fusional relationships*, referring to relationships in which the self merges with another. Turkle argues that girls tend to have such relationships because their original fusional relationship, with their mothers, was never broken up by the complications of the Oedipal stage and intervening fathers and so on.

If the term *fusional relationship* was used only by the psychoanalytically minded, I would not call attention to it here. But Turkle uses it in other contexts as well, such as when she argues that girls (and later, women) tend to have fusional relationships with objects as well as with people. Turkle notes how nine-year-old Anne tends to anthropomorphize the computer, treating it like a person; when she programs the computer she likes to project herself into the very space on the screen. In doing so, she reduces the distance between subject and object, which is another example of a fusional relationship.

We are accustomed to thinking of true science as neutral and objective, unaffected by the gender of the person who undertakes its practice. But some have recently rejected this view and instead chosen to speak of what Evelyn Keller has called "the genderization of science," meaning that there are at least two, equally valid scientific approaches, one "male" and one "female," but only the male mode, with its distinct separation of subject and object, has received the sanction and blessing of the official custodians of science.

Turkle's portrayal of the girl students' soft style of programming mastery is clearly not intended to be read as a manifestation of the hoary law that "biology is destiny." She is careful to note that she saw exceptions: girls that were hards and boys that were softs. But general patterns did tend to follow the lines of gender. If her observations hold true, perhaps we can use her findings to discover new ways to reduce lingering disparities in the achievements of girl and boy students in mathematics and the sciences. Here we must entertain the possibility that in the classroom of the future the technology of the personal computer may prove to be more liberating for the softs than the hards, for it is they who would be most likely to remain separated from the world of math and science. Turkle suggests that the softs can be just as competent as the hards, but they take a different route to get to the same destination.

CHILD PROGRAMMERS: THE FIRST GENERATION

Consider Robin, a four-year-old with blond hair and a pinafore, standing in front of a computer console, typing at its keyboard. She is a student at a nursery school that is introducing computers to very young children. She is playing a game that allows her to build stick figures by commanding the computer to make components appear and move into a desired position. The machine responds to Robin's commands and tells her when it does not understand an instruction. Many people find this scene disturbing. First, Robin is "plugged into" a machine. We speak of television as a "plug-in drug," but perhaps the very passivity of what we do with television reassures us. We are concerned about children glued to screens, but, despite what we have heard of Marshall McLuhan and the idea that "the medium is the message," the passivity of television encourages many of us to situate our sense of its impact at the level of the content of television programming. Is it violent or sexually suggestive? Is it educational? But Robin is not "watching" anything on the computer. She is manipulating—perhaps more problematic, *interacting with*—a complex technological medium. And the degree and intensity of her involvement suggests that (like the children at the video games) it is the medium itself and not the content of a particular

---

SOURCE: From *The Second Self* by Sherry Turkle. Copyright © 1984 by Sherry Turkle. Reprinted by permission of Simon & Schuster, Inc.

program that produces the more powerful effect. But beyond any specific fear, so young a child at a computer conflicts with our ideal image of childhood. The "natural" child is out of doors; machines are indoors. The natural child runs free; machines control and constrain. Machines and children don't go together.

Something else feels discordant, and that is the nature of Robin's interaction with the computer. She is not manipulating the machine by turning knobs or pressing buttons. She is writing messages to it by spelling out instructions letter by letter. Her painfully slow typing seems laborious to adults, but she carries on with an absorption that makes it clear that time has lost its meaning for her. Computers bring writing within the scope of what very young children can do. It is far easier to press keys on a keyboard than to control a pencil. Electronic keyboards can be made sensitive to the lightest touch; more important, they permit instant erasure. The computer is a forgiving writing instrument, much easier to use than even an electric typewriter.

That a four-year-old or a three-year-old might learn to make a fire poses a real physical danger, but it does not call anything about childhood into question. We find it easy to accept, indeed we are proud, when children develop physical skills or the ability to manipulate concrete materials earlier than we expect. But a basic change in the child's manipulation of symbolic materials threatens something deep. Central to our notion of childhood is the idea that children of Robin's age and younger speak but do not write.

Many people are excited by the possibility that writing may be brought within the range of capabilities of very young children. But others seem to feel that setting a four-year-old to writing does violence to a natural process of unfolding. For them, what is most disturbing about Robin is not her relationship to the machine, but her relationship to writing, to the abstract, to the symbolic. Opening the question of children and writing provokes a reaction whose force recalls that evoked by Freud's challenge to the sexual innocence of the child.

In the eighteenth century, Jean-Jacques Rousseau associated writing with moral danger in the most direct way. He saw the passage from nature to culture as the end of a community of free, spontaneous communication. Writing marked the point of rupture. In Rousseau's mind, this story of loss of community and communication projects itself onto the life of each individual. Each growing up is a loss of innocence and immediacy, and the act of writing symbolizes that the loss has taken place. To a certain extent, each of us reenacts the fall. Our first marks of pen on paper retrace the introduction of a barrier between ourselves and other people. Childhood, innocence, is the state of not writing.

The computer has become the new cultural symbol of the things that Rousseau feared from the pen: loss of direct contact with other people, the construction of a private world, a flight from real things to their representations. With programming, as for so many other things, the computer presence takes what was already a concern and gives it new form and new degree. If our ideas about childhood are called into question by child writers, what of child programmers? If childhood innocence is eroded by writing, how much more so by programming?

What happens when young children, grade-school children, become programmers? Faced with the reality of child experts who have appropriated the computers that dot grade schools and junior highs across the country, there is talk of a "computer generation" and of a new generation gap.

Sarah, a thirty-five-year-old lawyer and mother of three, feels an unbridgeable gap between herself and her son, and she alternates between agitation and resignation:

> I could have learned that "new math." I could understand, respect my son if his values turn out to be different than mine. I mean, I think I could handle the kinds of things that came up between parents and kids in the sixties. I would have talked to my son; I would have tried to understand. But my ten-year-old is into programming, into computers, and I feel that this makes his mind work in a whole different way.

Do computers change the way children think? Do they open children's minds or do they dangerously narrow their experience, making their thinking more linear and less intuitive? There is a temptation to look for a universal, isolable effect, the sort that still eludes experts on the effect of television.

The problem here is the search for a universal effect. I have found that different children are touched in remarkably different ways by their experience with the computer. However, by looking closely at how individual children appropriate the computer we can build ways to think about how the computer enters into development, and we begin to get some answers to our questions. In a sense, I turn the usual question around: instead of asking what the computer does to children I ask what children, and more important, what different kinds of children make of the computer. . . .

A Children's Computer Culture

When children learn to program, one of their favorite areas of work is computer graphics—programming the machine to place displays on the screen. The Logo graphics system available at Austen was relatively powerful. It provided thirty-two computational objects called sprites that appear on the screen when commanded to do so. Each sprite has a number. When called by its number and given a color and shape, it comes onto the screen with that shape and color: a red truck, a blue ball, a green airplane. Children can manipulate one sprite at a time, several of them, or all of them at once, depending on the effect they want to achieve. The sprites can take predefined shapes, such as trucks and airplanes, or they can be given new shapes designed on a special grid, a sprite "scratchpad." They can be given a speed and a direction and be set in motion. The direction is usually specified in terms of a heading from 0 to 360, where 0 would point the sprite due north, 90 would point it due east, 180 south, 270 west, and 360 north again.

At the time the system was introduced, the teachers thought the manipulation of headings would be too complex for second graders because it involves the concept of angles, so these children were introduced to the commands for making sprites appear, giving them shapes and colors, and placing them on the screen, but not for setting them in motion. Motion would be saved for later grades.

The curriculum held for two weeks. That is, it held until one second grader, Gary, caught on to the fact that something exciting was happening on the older children's screens, and knew enough to pick up the trick from a proud and talkative third grader. In one sense, the teachers were right: Gary didn't understand that what he was dealing with were "angles." He didn't have to. He wanted to make the computer do something, and he found a way to assimilate the concept of angle to something he already knew—secret codes. "The sprites have secret codes, like 10, 100, 55. And if you give them their codes they go in different directions. I've taught the code to fourteen second graders," he confided to a visitor. "We're sort of keeping it a secret. The teachers don't know. We haven't figured out all the codes yet, but we're working on it." Two weeks later, Gary and his friends were still cracking the code. "We're still not sure about the big numbers" (sprites interpret 361 as 1, one full revolution plus 1), but they were feeling very pleased with themselves.

Gary's discovery, not the only one of its kind, contributed to creating a general pattern at Austen. Students felt that computer knowledge belonged to them and not only to the teachers. Once knowledge had become forbidden fruit, once appropriation of it had become a personal challenge, teachers could no longer maintain their position as the rationers of "curricular materials." In a setting like Austen, ideas about programming travel the way ideas travel in active, dynamic cultures. They sweep through, carried by children who discover something, often by chance, through playful exploration of the machine.

Gary and his fellow decoders finally presented their discoveries to the authorities with pride of authorship. At Austen programming tricks and completed programs are valued—they are traded and they become gifts. In traditional school settings, finished book reports are presented to teachers who try to instill a sense of the class as community by asking the children to read them aloud to the group. In the context of children and programming projects, the sharing usually happens naturally. Children can't do much with each other's book reports, but they can do a great deal with each other's programs. Another child's program can be changed, new features can be added, it can be personalized. (One child can figure out how to get the computer to engage in a "dialogue," but a second child can change the script; one child can figure out how to write a program that will display an animated drawing of a rocket going to the moon, but a second child can build on it and have the rocket orbit once it gets there.) Most objects can't be given away and kept at the same time. But computer programs are easily shared, copied from one child's personal storage disk or cassette to that of another. As the child experiences it, the originator of the program gets to be famous. And other people get to build on his or her ideas.

Anne, an artistic fourth grader, had originated a program in which birds made of sprites fly across the sky and disappear behind clouds. One morning as we spoke, she glanced around the classroom. Five of the eight computers within view had objects disappearing, melting, and fading into other colors. "It's like a game of telephone," she said. "You start it, but then it changes. But you can always sort of see part of your original idea. And people know that you were the first."

At Austen we are faced with the growth of an intellectual community that we do not normally see among schoolchildren. What makes the community most special is that it includes children with a wide range of personalities, interests, and learning styles who express their differences through their styles of programming.

Jeff and Kevin

Jeff, a fourth grader, has a reputation as one of the school's computer experts. He is meticulous in his study habits, does superlative work in all subjects. His teachers were not surprised to see him excelling in programming. Jeff approaches the machine with determination and the need to be in control, the way he approaches both his schoolwork and his extracurricular activities. He likes to be, and often is, chairman of student committees. At the moment, his preoccupation with computers is intense: "They're the biggest thing in my life right now." He speaks very fast, and when he talks about his programs he speaks even faster, tending to monologue. He answers a question about what his program does by tossing off lines of computer code that for him seem to come as naturally as English. His typing is expert—he does not look at the code as it appears on the screen. He conveys the feeling that he is speaking directly to an entity inside. "When I program I put myself in the place of the sprite. And I make it do things."

Jeff is the author of one of the first space-shuttle programs. He does it, as he does most other things, by making a plan. There will be a rocket, boosters, a trip through the stars, a landing. He conceives the program globally; then he breaks it up into manageable pieces. "I wrote out the parts on a big piece of cardboard. I saw the whole thing in my mind just in one night, and I couldn't wait to come to school to make it work." Computer scientists will recognize this global "top-down," "divide-and-conquer" strategy as "good programming style." And we all recognize in Jeff someone who conforms to our stereotype of a "computer person" or an engineer—someone who would be good with machines, good at science, someone organized, who approaches the world of things with confidence and sure intent, with the determination to make it work.

Kevin is a very different sort of child. Where Jeff is precise in all of his actions, Kevin is dreamy and impressionistic. Where Jeff tends to try to impose his ideas on other children, Kevin's warmth, easygoing nature, and interest in others make him popular. Meetings with Kevin were often interrupted by his being called out to rehearse for a school play. The play was *Cinderella*, and he had been given the role of Prince Charming. Kevin comes from a military family; his father and grandfather were both in the Air Force. But Kevin has no intention of following in their footsteps. "I don't want to be an army man. I don't want to be a fighting man. You can get killed." Kevin doesn't like fighting or competition in general. "You can avoid fights. I never get anybody mad—I mean, I try not to."

Jeff has been playing with machines all his life—Tinkertoys, motors, bikes—but Kevin has never played with machines. He likes stories, he likes to read, he is proud of knowing the names of "a lot of different trees." He is artistic and introspective. When Jeff is asked questions about his activities, about what he thinks is fun, he answers in terms of how to do them right and how well he does them. He talks about video games by describing his strategy breakthroughs on the new version of Space Invaders: "Much harder, much trickier than the first one." By contrast, Kevin talks about experiences in terms of how they make him feel. Video games make him feel nervous, he says. "The computer is better," he adds. "It's easier. You get more relaxed. You're not being bombarded with stuff all the time."

Kevin too is making a space scene. But the way he goes about it is not at all like Jeff's approach. Jeff doesn't care too much about the detail of the form of his rocket ship; what is important is getting a complex system to work together as a whole. But Kevin cares more about the aesthetics of the graphics. He spends a lot of time on the shape of his rocket. He abandons his original idea ("It didn't look right against the stars") but continues to "doodle" with the scratchpad shape-maker. He works without plan, experimenting, throwing different shapes onto the screen. He frequently stands back to inspect his work, looking at it from different angles, finally settling on a red shape against a black night— a streamlined, futuristic design. He is excited and calls over two friends. One admires the red on the black. The other says that the red shape "looks like fire." Jeff happens to pass Kevin's machine on the way to lunch and automatically checks out its screen, since he is always looking for new tricks to add to his toolkit for building programs. He shrugs. "That's been done." Nothing new there, nothing technically different, just a red blob.

Everyone goes away and Kevin continues, now completely taken up by the idea that the red looks like fire. He decides to make the ship white so that a red shape can be red fire "at the bottom." A long time is spent making the new red fireball, finding ways to give it spikes. And a long time is spent adding detail to the now white ship. With the change of color, new possibilities emerge: "More things will show up on it." Insignias, stripes, windows, and the project about which Kevin is most enthusiastic: "It can have a little seat for the astronaut." When Jeff programs he puts himself in the place of the sprite; he thinks of himself as an abstract computational object. Kevin says that, as he works, "I think of myself as the man inside the rocket ship. I daydream about it. I'd like to go to the moon."

By the next day Kevin has a rocket with red fire at the bottom. "Now I guess I should make it move . . . moving and wings . . . it should have moving and wings." The wings turn out to be easy, just some more experimenting with the scratchpad. But he is less certain about how to get the moving right.

Kevin knows how to write programs, but his programs emerge— he is not concerned with imposing his will on the machine. He is concerned primarily with creating exciting visual effects and allows himself to be led by the effects he produces. Since he lets his plans change as new ideas turn up, his work has not been systematic. And he often loses track of things. Kevin has lovingly worked on creating the rocket, the flare, and a background of twinkling stars. Now he wants the stars to stay in place and the rocket and the flare to move through them together.

It is easy to set sprites in motion: just command them to an initial position and give them a speed and a direction. But Kevin's rocket and red flare are two separate objects (each shape is carried by a different sprite) and they have to be commanded to move together at the same speed, even though they will be starting from different places. To do this successfully, you have to think about coordinates and you have to make sure that the objects are identified differently so that code for commanding their movement can be addressed to each of them independently. Without a master plan Kevin gets confused about the code numbers he has assigned to the different parts of his program, and the flare doesn't stay with the rocket but flies off with the stars. It takes a lot of time to get the flare and the ship back together. When Jeff makes a mistake, he is annoyed, calls himself "stupid," and rushes to correct his technical error. But when Kevin makes an error, although it frustrates him he doesn't seem to resent it. He sometimes throws his arms up in exasperation: "Oh no, oh no. What did I do?" His fascination with his effect keeps him at it.

In correcting his error, Kevin explores the system, discovering new special effects as he goes along. In fact, the "mistake" leads him to a new idea: the flare shouldn't go off with the stars but should drop off the rocket, "and then the rocket could float in the stars." More experimenting, trying out of different colors, with different placements of the ship and the flare. He adds a moon, some planets. He tries out different trajectories for the rocket ship, different headings, and different speeds; more mistakes, more standing back and admiring his evolving canvas. By the end of the week Kevin too has programmed a space scene.

Styles of Mastery

Jeff and Kevin represent cultural extremes. Some children are at home with the manipulation of formal objects, while others develop their ideas more impressionistically, with language or visual images, with attention to such hard-to-formalize aspects of the world as feeling, color, sound, and personal rapport. Scientific and technical fields are usually seen as the natural home for people like Jeff; the arts and humanities seem to belong to the Kevins.

Watching Kevin and Jeff programming the same computer shows us two very different children succeeding at the same thing—and here it must be said that Kevin not only succeeded in creating a space scene, but, like Jeff, he learned a great deal about computer programming and mathematics, about manipulating angles, shapes, rates, and coordinates. But although succeeding at the same thing, they are not doing it the same way. Each child developed a distinctive style of mastery— styles that can be called hard and soft mastery.

Hard mastery is the imposition of will over the machine through the implementation of a plan. A program is the instrument of premeditated control. Getting the program to work is more like getting "to say one's piece" than allowing ideas to emerge in the give-and-take of conversation. The details of the specific program obviously need to be "debugged"—there has to be room for change, for some degree of flexibility in order to get it right—but the goal is always getting the program to realize the plan.

Soft mastery is more interactive. Kevin is like a painter who stands back between brushstrokes, looks at the canvas, and only from this contemplation decides what to do next. Hard mastery is the mastery of the planner, the engineer, soft mastery is the mastery of the artist: try this, wait for a response, try something else, let the overall shape emerge from an interaction with the medium. It is more like a conversation than a monologue.

Hard and soft mastery recalls anthropologist Claude Lévi-Strauss' discussion of the scientist and the *bricoleur*. Lévi-Strauss used the term *bricolage*, tinkering, to make a distinction between Western science and the science of preliterate societies. The former is a science of the abstract, the latter is a science of the concrete. Like the *bricoleur*, the soft master works with a set of concrete elements. While the hard master thinks in terms of global abstractions, the soft master works on a problem by arranging and rearranging these elements, working through new combinations. Although the *bricoleur* works with a closed set of materials, the results of combining elements can lead to new and surprising results.

Mastery and Personality

Computer programming is usually thought of as an activity that imposes its style on the programmer. And that style is usually presumed to be closer to Jeff and his structured, "planner's" approach than to Kevin and his open, interactive one. In practice, computer programming allows for radical differences in style. And looking more closely at Jeff and Kevin makes it apparent that a style of dealing with the computer is of a piece with other things about the person—his or her way of facing the world, of coping with problems, of defending against what is felt as dangerous. Programming style is an expression of personality style.

For example, the hard masters tend to see the world as something to be brought under control. They place little stock in fate. In child's terms, they don't believe much in a rabbit's foot or a lucky day. Jeff is popular and sociable, but he likes to be committee chairman, the one who controls the meeting. From the earliest ages most of these children have preferred to operate on the manipulable—on blocks, on Tinkertoys, on mechanisms. It is not surprising that the "hards" sometimes have more difficulty with the give-and-take of the playground. When your needs for control are too great, relationships with people become tense and strained. The computer offers a "next-best" gratification. The Tinkertoy is inert. The computer is responsive. Some children even feel that when they master it they are dominating something that "fights back." It is not surprising that hard masters take avidly to the computer. It is also not surprising that their style of working with the computer emphasizes the imposition of will.

The soft masters are more likely to see the world as something they need to accommodate to, something beyond their direct control. In general, these children have played not with model trains and Erector sets but with toy soldiers or with dolls. They have taken the props (cowboy hats, guns, and grownup clothes for dress-up) from the adult world and used them in fantasy play with other children. In doing so, they have learned how to negotiate, compromise, empathize. They tend to feel more impinged upon, more reactive. As we have seen, this accommodating style is expressed in their relational attitude toward programming as well as in their relationships with people. . . .

Mastery and Gender

I have used boys as examples in order to describe hard and soft mastery without reference to gender. But now it is time to state what might be anticipated by many readers: girls tend to be soft masters, while the hard masters are overwhelmingly male. At Austen, girls are trying to forge relationships with the computer that bypass objectivity altogether. They tend to see computational objects as sensuous and tactile and relate to the computer's formal system not as a set of unforgiving "rules," but as a language for communicating with, negotiating with, a behaving, psychological entity.

There are many reasons why we are not surprised that girls tend to be soft masters. In our culture girls are taught the characteristics of soft mastery—negotiation, compromise, give-and-take—as psychological virtues, while models of male behavior stress decisiveness and the imposition of will. Boys and girls are encouraged to adopt these stances in the world of people. It is not surprising that they show up when children deal with the world of things. The girl child plays with dolls, imagined not as objects to command but as children to nurture. When the boy unwraps his birthday presents they are most likely to be Tinkertoys, blocks, Erector sets—all of which put him in the role of builder.

Thinking in terms of dolls and Erector sets, like talking about teaching negotiation and control, suggests that gender differentiation is a product of the social construction that determines what toys and what models of correct behavior are given to children of each sex. Psychoanalytic thought suggests many ways in which far earlier processes could have their role to play; styles of mastery may also be rooted in the child's earliest experiences. One school of thought, usually referred to as "object relations theory," is particularly rich in images that suggest a relation between styles of mastery and gender differences.

It portrays the infant beginning life in a closely bonded relationship with the mother, one in which boundaries between self and other are not clear. Nor does the child experience a separation between the self and the outer world. The gradual development of a consciousness of separate existence begins with a separation from the mother. It is fraught with conflict. On the one hand, there is a desire to return to the comfort of the lost state of oneness. On the other hand, there is the pleasure of autonomy, of acting on independent desire. Slowly the infant develops the sense of an "objective" reality "out there" and separate from the self. Recently, there has been serious consideration of the ways in which this process may take on a sense of gender. Since our earliest and most compelling experiences of merging are with the mother, experiences where boundaries are not clear become something "female." Differentiation and delineation, first worked through in a separation from the mother, are marked as "not-mother," not-female.

Up to this point the experiences are common to girls and boys. But at the Oedipal stage, there is a fork in the road. The boy is involved in a fantasized romance with the mother. The father steps in to break it up and, in doing so, strikes another blow against fusional relationships. It is also another chance to see the pressure for separation as male. This is reinforced by the fact that this time the boy gives up the idea of a romance with the mother through identifying himself with his father. Thus, for the boy, separation from the mother is more brutal, because in a certain sense it happens twice: first in the loss of the original bonded relationship, then again at the point of the Oedipal struggle.

Since separation from the mother made possible the first experiences of the world as "out there," we might call it the discovery of the "objective." Because the boy goes through this separation twice, for him objectivity becomes more highly charged. Boys feel a greater desire for it: the objective, distanced relationship feels like safe, approved ground. There is more of a taboo on the fusional, along with a correspondingly greater fear of giving in to its forbidden pleasures. According to this theory the girl is less driven to objectivity because she is allowed to maintain more elements of the old fusional relationship with the mother, and, correspondingly, it is easier for her to play with the pleasures of closeness with other objects as well.

Anne and Mary

In the eyes of a true hard programmer like Jeff, his classmate Anne, also nine, is an enigma. On the one hand, she hardly seems serious about the computer. She is willing to spend days creating shimmering patterns on the screen in a kind of "moiré effect" and she doesn't seem to care whether she gets her visual effects with what Jeff would classify as technically uninteresting "tricks" or with what he would see as "really interesting" methods. Jeff knows that all the children anthropomorphize the computer to a certain extent; everyone says things like "My program knows how to do this" or "You have to tell the computer what speed you want the sprites to go," but Anne carries anthropomorphizing to what, to Jeff, seems like extreme lengths. For example, she insists on calling the computer "he," with the explanation "It doesn't seem right to call it an it." All the same, this doesn't keep her from getting down to serious programming. She has made some technical inventions, and Jeff and the other male hard masters recognize that if they want to keep abreast of the state of the art at Austen they must pay attention to what Anne is doing. And Anne knows how to take advantage of her achievements. She analogizes the spread of programming ideas to the game of telephone and enjoys seeing versions of her ideas on half a dozen screens. "They didn't copy me exactly, but I can recognize my idea." Jeff's grudging acknowledgment of Anne's "not quite serious" accomplishments seems almost a microcosm of reactions to competent women in society as a whole. There, as at Austen, there is appreciation, incomprehension, and ambivalence.

When Jeff talks with the other male experts about the computer, they usually talk "shop" about technical details. Anne, on the other hand, likes to discuss her strong views about the machine's psychology. She has no doubt that computers have psychologies: they "think," as people do, although they "can't really have emotions." Nevertheless, the computer might have preferences. "He would like it if you did a pretty program." When it comes to technical things, she assumes the computer has an aesthetic: "I don't know if he would rather have the program be very complicated or very simple."

Anne thinks about whether the computer is alive. She says that the computer is "certainly not alive like a cat," but it is "sort of alive," it has "alive things." Her evidence comes from the machine's responsive behavior. As she types her instructions into the machine, she comments, "You see, this computer is close to being alive because he does what you are saying."

This remark is reminiscent of the talk among the somewhat younger children who were preoccupied with sorting out the computer's status as a living or a not living thing. There is, however, a difference. For the younger children, these questions have a certain theoretical urgency. For Anne, they are both less urgent and part of a practical philosophy: she has woven this way of seeing the computer into her style of technical mastery.

Anne wants to know how her programs work and to understand her failures when they don't. But she draws the line between understanding and not understanding in a way that is different from most of the hard-master boys at a similar degree of competence. For them, a program (like anything else built out of the elements of a formal system) is either right or wrong. Programs that are correct in their general structure are not "really correct" until the small errors, the bugs, are removed.

For a hard programmer like Jeff, the bugs are there to ferret out. Anne, on the other hand, makes no demand that her programs be perfect. To a certain degree, although to put it too flatly would be an exaggeration, when she programs the computer she treats it as a person. People can be understood only incompletely: because of their complexity, you can expect to understand them only enough to get along, as well as possible for maintaining the kind of relationship you want. And when you want people to do something, you don't insist that it be done exactly as you want it, but only "near enough." Anne allows a certain amount of negotiation with the computer about just what should be an acceptable program. For her, the machine is enough alive to deserve a compromise.

This "negotiating" and "relational" style is pervasive in Anne's work but is more easily described by an example from her classmate Mary, another soft-mastery programmer and an even stauncher lobbyist for the use of personal pronouns to refer to computers. Mary differs strikingly from Anne in having a soft style that is verbal where Anne's is consistently visual.

Mary wanted to add a few lines of dialogue to the end of a game program. Her original idea was that the computer would ask the player, "Do you want to play another game?" If the player typed "Yes," a new game would start. If the player typed "No," the machine would print out the final score and "exit" the program—that is, put the machine back into a state where it is ready for anything, back to "top level." She writes a program that has two steps, captured in the following English-language rendition of the relevant Logo instructions:

If what-the-user-types is "Yes," start a new game.
If what-the-user-types is "No," print score and stop.

As instructions to an intelligent person, these two statements are unambiguous. Not so as instructions to a computer. The program "runs," but not quite as Mary originally planned. The answer "Yes" produces the "right" behavior, a new game. But in order to get the final score and exit, it is necessary to type "No" twice. Mary knew this meant there was an "error," but she liked this bug. She saw the behavior as a humanlike quirk.

What was behind the quirk? The computer is a serial machine; it executes each instruction independently. It gets up to the first instruction that tells it to wait until the user types something. If this something is "Yes," a new game is started up. If the user doesn't type "Yes," if, for example, he or she types "No," the computer does nothing except pass on to the next instruction without "remembering" what has come before. The second instruction, like the first, tells the computer to wait until the user has typed something. And if this something is "No," to print the score and stop.

Now the role of the two "Nos" is clear. A single "No" will leave the computer trying to obey the second instruction—that is, waiting for the user to type something. There are ways of fixing this bug, but what is important here is the difference in attitude between a programmer like Jeff, who would not rest until he fixed it, and a programmer like Mary, who could figure out how to fix it but decides not to. Mary *likes* this bug because it makes the machine appear to have more of a personality. It lets you feel closer to it. As Mary puts it, "He will not take no for an answer" unless you really insist. She allows the computer its idiosyncrasies and happily goes on to another program.

Mary's work is marked by her interest in language. Anne's is equally marked by her hobby, painting. She uses visual materials to create strategies for feeling "close to the machine."

Anne had become an expert at writing programs to produce visual effects of appearance and disappearance. In one, a flock of birds flies through the sky, disappears at the horizon and reappears some other place and time. If all the birds are the same color, such as red, then disappearance and appearance could be produced by the commands "SETCOLOR :INVISIBLE" to get rid of them and "SETCOLOR :RED" to make them appear. But since Anne wants the birds to have different colors, the problem of the birds reappearing with their original color is more complicated.

There is a classical method for getting this done: get the program to "store away" each bird's original color before changing that color to "invisible," and then to recall the color when the birds are to reappear. This method calls for an algebraic style of thinking. You have to think about variables and use a variable for each bird—for example, letting A equal the color of the first bird, B the color of the second bird, and so on. Anne will use this kind of method when she has to, but she prefers another kind, a method of her own invention that has a different feel.

She likes to feel that she is there among her birds, manipulating them much in the way she can manipulate physical materials. When you want to hide something on a canvas, you paint it out, you cover it with something that looks like the background. This is Anne's solution. She lets each bird keep its color, but she makes her program "hide it" by placing a screen over it. She designs a sprite that will screen the bird when she doesn't want it seen, a sky-colored screen that makes it disappear. Just as the computer can be programmed to make a bird-shaped object on the screen, it can be programmed to make an opaque sky-colored square act as a screen.

Anne is programming a computer, but she is thinking like a painter. She is not thinking about sprites and variables. She is thinking about birds and screens. Anne's way of making birds appear and disappear doesn't make things technically easy. On the contrary, to maintain her programming aesthetic requires technical sophistication and ingenuity.

For example, how does the program "know" where the bird is so as to place the screen on it? Anne attaches the screen to the bird when the bird is created, instead of putting it on later. The screen is on top of the bird at all times and moves with the bird wherever it goes. Thus she has invented a new kind of object, a "screened bird." When Anne wants the bird to be seen, the screen is given the "invisible" color, so the bird, whatever its color, shows right through it. When she wants the bird to disappear, the screen is given the color of the sky. The problem of the multiplicity of bird colors is solved. A bird can have any color. But the screens need only two colors, invisible or sky blue. A bird gets to keep its color at all times. It is only the color of its screen that changes. The problem of remembering the color of a particular bird and reassigning it at a particular time has been bypassed.

Anne's bird program is particularly ingenious, but its programming style is characteristic of many of the girls in her class. Most of the boys seem driven by the pleasures of mastering and manipulating a formal system: for them, the operations, the programming instructions, are what it is all about. For Anne, pleasure comes from being able to put herself into the space of the birds. Her method of manipulating screens and birds allows her to feel that these objects are close, not distant and untouchable things that need designation by variables. The ambivalence of the computational sprite—an object at once physical and abstract— allows it to be picked up differently by hards and softs. Anne responds to the sprites as physical objects. Her work with them is intimate and direct. The formal operations need to be mastered, but they are not what drive her.

No one would find Anne's relation to the birds and the screens surprising if it were in the context of painting or making collages with scraps of this and that. There we expect to find "closeness to the object." But finding a sensual aesthetic in the development of a computer program surprises us. We tend to think of programming as the manipulation of a formal system which, like the objects for scientific inquiry, is "out there" to be operated on as something radically split from the self.

Gender and Science

Evelyn Keller has coined the phrase "the genderization of science." She argues that what our culture defines as the scientific stance toward the world corresponds to the kind of relationships with the object world that most men (if we follow psychoanalytic theories of development) would be expected to find most comfortable. It is a relationship that cuts off subject from object.

Scientific objects are placed in a "space" psychologically far away from the world of everyday life, from the world of emotion and relationships. Men seem able, willing, and invested in constructing these separate "objective" worlds, which they can visit as neutral observers. In this way the scientific tradition that takes objectivity as its hallmark is also defined as a male preserve. Taking it from the other side, we can see why men would be drawn to this construction of science. Men are highly invested in objective relationships with the world. Their earliest experiences have left them with a sense of the fusional as taboo, as something to be defended against. Science, which represents itself as revealing a reality in which subject and object are radically separated, is reassuring. We can also see why women might experience a conflict between this construction of science and what feels like "their way" of dealing with the world, a way that leaves more room for continuous relationships between self and other. Keller adds that the presentation of science as an extreme form of objective thinking has been reinforced by the way in which male scientists traditionally write and speak about their work. A characterization of science that appears to "gratify particular emotional needs" may "give rise to a self-selection of scientists—a self-selection which would in turn lead to a perpetuation of that characterization."

In Anne's classroom, nine- and ten-year-old girls are just beginning to program. The fact that they relate to computational objects differently from boys raises the question of whether with growing expertise they will maintain their style or whether we are simply seeing them at an early stage before they become "recuperated" into a more objective computational culture. In my observation, with greater experience soft masters, male and female, reap the benefits of their long explorations, so that they appear more decisive and more like "planners" when they program on familiar terrains. But the "negotiating" and "relational" style remains behind the appearance and resurfaces when they tackle something new.

Lorraine is the only woman on a large team working on the design of a new programming language. She expresses her sense of difference with some embarrassment.

> I know that the guys I work with think I am crazy. But we will be working on a big program and I'll have a dream about what the program feels like inside and somehow the dream will help me through.
>
> When I work on the system I know that to everybody else it looks like I'm doing what everyone else is doing, but I'm doing that part with only a small part of my mind. The rest of me is imagining what the components feel like. It's like doing my pottery. . . . Keep this anonymous. I know this sounds stupid.

Shelley is a graduate student in computer science who corrects me sharply when I ask her when she got interested in electronics and ma-

chines. "Machines," she responds, "I am definitely not into machines."
And she is even less involved with electronics:

> My father was an electrician and he had all of these machines around. All
> of these wires, all of this stuff. And he taught my brothers all about it. But
> all I remember him telling me was, "Don't touch it, you'll get a shock." I
> hate machines. But I don't think of computers as machines. I think of
> moving pieces of language around. Not like making a poem, the way you
> would usually think of moving language around, more like making a piece
> of language sculpture.

These words are reminiscent of women in other scientific disciplines. Barbara McClintock, an eminent biologist, describes her work as
an ongoing "conversation" with her materials, and she speaks of frustration with the way science is usually done: "If you'd only just let the
materials speak to you . . ." In an interview with her biographer, Evelyn
Keller, McClintock described her studies of neurospora chromosomes
(so small that others had been unable to identify them) in terms that
recall Anne's relationship with the birds and the screens. "The more
she worked with the chromosomes, the bigger they got; until finally, 'I
wasn't outside, I was down there—I was part of the system.' . . . As
'part of the system' even the internal parts of the chromosomes become
visible. 'I actually felt as if I were down there and these were my
friends.' "

Keller comments that McClintock's "fusion" with her objects of
study is something experienced by male scientists. But perhaps Mc-
Clintock was able to exploit this less distanced model of scientific
thought, far from the way science was discussed in the 1950s, more
fully, visibly, and less self-consciously, because she is a woman. This is
surely the case for the girls in the Austen classrooms. Their artistic,
interactive style is culturally sanctioned. Of course, with children, as in
the larger world, the lines of division are not rigid. Some girls are hard
masters and I purposely took a boy as the first case of a soft master—
Kevin, who did not see the sprites as "outside" but who is right there
with them, who imagines himself a traveler in the rocket ship, taking
himself and his daydreams with him.

Children working with computers are a microcosm for the larger
world of relations between gender and science. Jeff took the sprite as
an object apart and in a world of its own. When he entered the sprite
world, it was to command it better. Kevin used the sprite world to
fantasize in. Anne does something more. She moves further in the direction I am calling "feminine," further in the direction of seeing herself
as in the world of the sprite, further in the direction of seeing the sprite
as sensuous rather than abstract. When Anne puts herself into the sprite
world, she imagines herself to be a part of the system, playing with the
birds and the screens as though they were tactile materials.

Science is usually defined in the terms of the hard masters: it is the place for the abstract, the domain for a clear and distinct separation between subject and object. If we accept this definition, the Austen classroom, with its male hard masters, is a microcosm of the male genderization of science. But what about Anne and Mary? What about the other girls like them who are exploring and mastering the computer? Should we not say that they too are "little scientists"? If we do, then we see at Austen not only a model of the male model that characterizes "official science," but a model of how women, when given a chance, can find another way to think and talk about the mastery not simply of machines but of formal systems. And here the computer may have a special role. It provides an entry to formal systems that is more accessible to women. It can be negotiated with, it can be responded to, it can be psychologized.

ABOUT THE EDITOR

Randall E. Stross received his B.A. from Macalester College and his Ph.D. in history from Stanford University. He was on the faculty of the Colorado School of Mines from 1982 to 1986, where he taught courses on the modern history of technology and society. Since 1986, he has been an associate professor at San Jose State University, where he teaches courses on business and technology.

Stross's own research has focused on problems in the transfer abroad of American technology. While a graduate student, he lived and studied two years in the People's Republic of China, conducting research on the transfer of American agricultural technology to China in the early twentieth century. The resulting book, *The Stubborn Earth: American Agriculturalists on Chinese Soil, 1898-1937*, was published by the University of California Press (1986). At present, he is investigating more recent efforts to transfer American know-how to China, focusing on the attempted transfer of American management models over the course of the past decade.

A NOTE ON THE TYPE

The text of this book was set 10/12 Palatino using a film version of the face designed by Hermann Zapf that was first released in 1950 by Germany's Stempel Foundry. The face is named after Giovanni Battista Palatino, a famous penman of the sixteenth century. In its calligraphic quality, Palatino is reminiscent of the Italian Renaissance type designs, yet with its wide, open letters and unique proportions it still retains a modern feel. Palatino is considered one of the most important faces from one of Europe's most influential type designers.

Composed by Graphic World Inc., St. Louis, Missouri

Printed and bound by Malloy Lithographing, Inc., Ann Arbor, Michigan